Estradiol Receptors
In Benign and Malignant Ovarian Tumors

Prof. Dr. Sami A. AL-Mudhaffar
Dr. Ali Saad Elewi

ChapterOne

Introduction and Literature Survey

1.1. Steroid hormones

Under control of the anterior pituitary hormones, the ovaries, the testes, and the adrenals secrete a group of steroid hormones! t. :!!. The cyclopenteno phenanthrene ring structure is the basic carbon skeleton for all steroid hormones. The carbon skeleton consists of a five-carbon cyclopentane ring conjoined to a phenanthrene molecule. which is itself made up of three six-carbon nngs configured as shovvn in Figure (1.1).:;.[41].

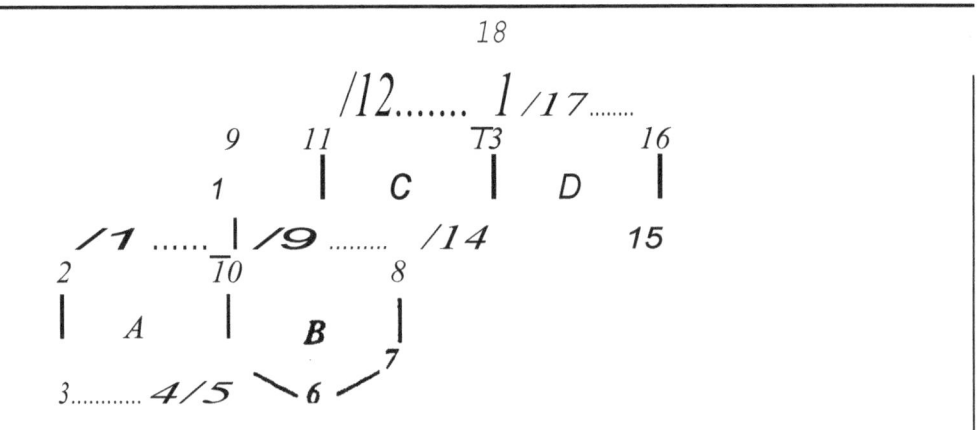

Steroid nucleus
(cyclopentenophenanthrene)

Figure (1.1)- Structure formula of the basic steroid nucleus[3] *.*

Cholesterol is the common precursor for all steroids and for this reason, the system of nomenclature relates to this steroid c[5]. The steroid hormones are classified depending on the number of carbon atoms in the molecule and their substituents. C-2 1 steroids include progesterone and the glucocorticoids. C-19 steroids include testosterone, dihydrotestosterone, and the adrenal androgens, and estrogens are in the C-18 group [11][6][7].

Steroid hormones are synthesized through a series of enzymatic reactions starting with a common precursor, cholesterol. The enzymes necessary to bring about specific changes in the cholesterol molecule are synthesized after stimulation by the pituitary hormones (I). When steroid hormones released into the circulation, the gonadal steroids bind to plasma proteins, and only unbound (free) steroids are biologically active <S>.

The steroid sex hormones enter cells by passive diffusion, bind to specific receptor proteins in the cytoplasm, and then move to the nucleus where they bind to specific sites on the chromosomes. These hormones exert their effects through gene-hormone interactions, which stimulate the synthesis of specific messenger ribonucleic acid [181].

Steroid inactivation occurs mainly in the liver, and the inactivated metabolites are excreted by the kidney. In general, biological activity is removed by saturation of the double bonds and the steroid is made water-soluble by conjugation with an acid and hence easier to secrete in the urine <[91].

1.1.I. Estrogen

1.1.1.1. Function

Estrogens are necessary for the normal physical maturation of the female. In concert with other hormones, estrogens provide for the reproductive processes of ovulation no', implantation, pregnancy, parturition, and lactation by stimulating the development and maintaining the growth of the accessory organs ([8,11]). Estrogens cause, development ofthe stromal tissues ofthe breasts, growth of an extensive ductile system, and deposition of fat in the breasts [112] • In [131] addition to their effects on the gro\vth of uterine muscle, estrogens play an important role in the development of the endometrial lining - [13]>. And have an effect on the mucosal lining of the fallopian tubes similar to that on the uterine endometrium <12).

Estrogens also have well established effects on plasma proteins that influence endocrine testing. They increase levels of SHBG, corticosteroid-binding globulin, and th)Toxine-binding globulin. Hence, boys and girls have comparable levels of SHBG, but adultmen have SHBG levels that are about one-half those of adult women [114, 15]. On the other hand, estradiol administered to chicks causes greatly increased rates of synthesis of characteristic egg proteins by the oviduct, particularly ovalbumin and ovovitellin. Thus estradiol prepares the ovary for egg formation [1161].

Moderate levels of estrogens in the blood inhibit both the release of GnRH by the hypothalamus and secretion of LH and FSH by the anterior pituitary s:!land [17-19>].

1.1.1.2. Chemistry

Estrogens are deriYatives of the parent hydrocarbon estrane. which is an 18-carbon molecule with an aromatic ring A and amethyl group at C-13. Estrogens possess a phenolic hydroxyl group at C-3, which gives the compounds acidic properties <[13]

Only three estrogens are present in significant quantities in the plasma of the human female: estradiol (E_1), estrone (EJ), and estriol (E_3) [11, 13, 22]. The formula for which are shov.·n in Figure (1-2) [112]. Ofthese, estradiol is the most biologically potent and the most abundantly secreted product of the ovary [18, 13]. Small amounts of estrone are also secreted, but most of this is formed in the peripheral tissues from androgens secreted by the adrenal and by the ovarian thecal cells. estriol is aweak estrogen, it is an oxidative product derived from both estradiol and estrone , the conversion occurring mainly in the liver [11=- 131]. The estrogenic potency of estradiol is 12 times that of estrone and 80 times that of estriol [1121].

Figure (1.2)- Chemical formulas of the estrogens ⁰².

1.1.1.3. Biosynthesis

Estrogens are secreted in acyclic pattern as a result of the interaction between the hypothalan:tic-releasing factors (GnRH) and the pituitary gonadotropic hormones (FSH and LH) <[8]>. In normal women, most estrogens are secreted by the ovary and, during pregnancy, by the placenta, adrenals and testes are also believed to secrete minute quantities of estrogens c[2 6 14 23 241].

Estrogen formation from androstenedione and testosterone is by a complex senes of enzymatic reactions collectively referred to as the aromatase complex its name is derived from the fact that it converts aliphatic to aromatic steroids. Aromatase requires androgen substrates for its action, hence androstenedione and testosterone are the precursors [5]. Estradiol is formed if the substrate of this enzyme complex is testosterone, whereas estrone results from the aromatization of androstenedione <[5 18]>, Figure (1.3)[1251]. In both cases the

oxygen group at position 17 remains unchanged during the conversion from the respective an^{dr}ogen precursor..[5].- ...

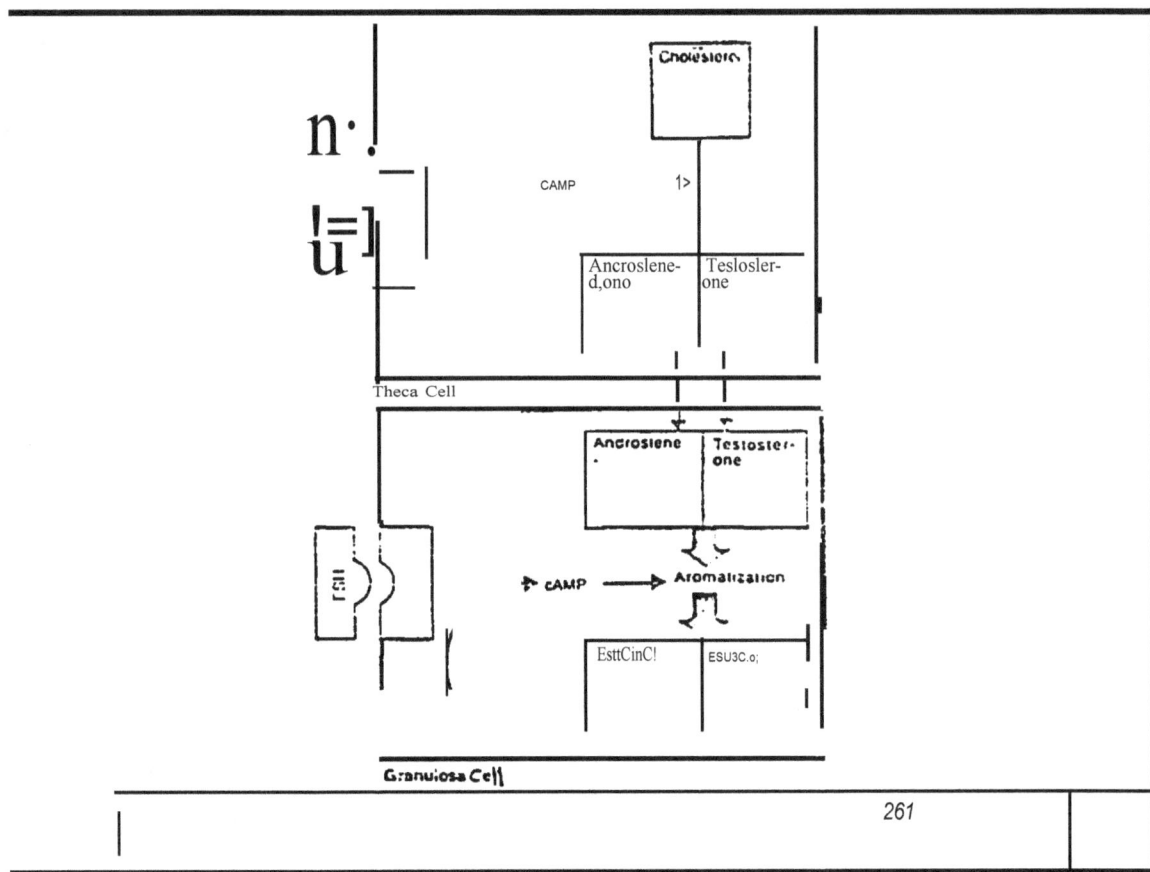

1.1.1.4. Transport

Greater than 97% of circulating estradiol is bound to plasma protein. It is bound specifically and with high affinity to SHBG and nonspecifically to albumin. SHBG concentrations are increased by estrogen and therefor are higher in women than in men. They are also increased during pregnancy and oral contraceptive use cs. ı-t [27]. SHBG concentrations may decrease in obesity, or androgen excess. In women, estradiol circulates bound 40%-60% to SHBG and 40%-60% to albumin. SHBG has a higher affinity for testosterone than estradiol, therefor, in men, estradiol circulates 20%-30% bound to SHBG and 70o/o-80% bound to albumin [1] ı-t [2]ıs'.

The binding of estradiol to its carrier protein is a reversible process resulting in an equilibrium between bound and free fractions (1). It is important to understand that, only unbound (free) estrogens are biologically active. Therefore, the function of plasma binding protein may be to provide a pool of circulating hormones with delayed metabolic clearances and favorable water solubilities that increase or decrease the free pool as needed [s.2s].

1.1.1.5. Metabolis1n

As is the case for other steroids, the Ih·er is the pnmary site for the inactivation of estrogens. The main biochemical reactions are hydroxylation. oxidation, reduction, and methylation [4.22.29]. Ring A of the estrogens remains unaltered because of the chemical stability of aromatic ring systems [15]. Hydroxylation that occur at carbon< of ring A impart a catechol nature to the estrogen, and for this reason these estrogens are referred to as catechol estrogens. The phenolic hydroxy group at C-3 is almost conjugated with sulfate group, whereas hydroxylation at C-16 and with most hydroxy groups on ring D are usually conjugated with glucuronic acid [51].

The conjugated estrogens are water-soluble and do not bind to transport proteins, thus, they are excreted readily in the bile feces, and urine ns'.

Estrogen metabolism in postmenopausal women is more complex because of its indirect production through the peripheral metabolism of androgens as well as the conversion of estrone to estradiol and vice versa cJo. [321].

1.1.1.6. Mechanism o[action

The estrogen enters the target cell by diffusion through the cell membrane and is quickly bound by the specific receptor in the cytoplasm n. [5.33.35]. \\'hen an estrogen binds to its receptor it is generally thought to produce a conformational change in the receptor protein. Figure (1.3) this al1teration in the

binding protein seems to be a prerequisite for translocation of the estrogen-receptor complex to the cell nucleus <[33,35,38]>. Translocation of receptor from the cytoplasm to the nucleus, as a consequence of estrogen binding, is required for estrogen action. Translocation occurs rapidly, and it appears to be a temperature-dependent process <[34,36,39]>.

Within the nucleus the estrogen-receptor complex binds to a specific region of DN , called the estrogen response element, and this interaction leads to the formation of new messenger R.NA (mRNA). The mRNA then migrates into the cytoplasm, where it attaches to ribosomes. The mRNA is translated to affect the formation of proteins that eventually bring about specific changes in the traget cell <[19,29,34,40]>.

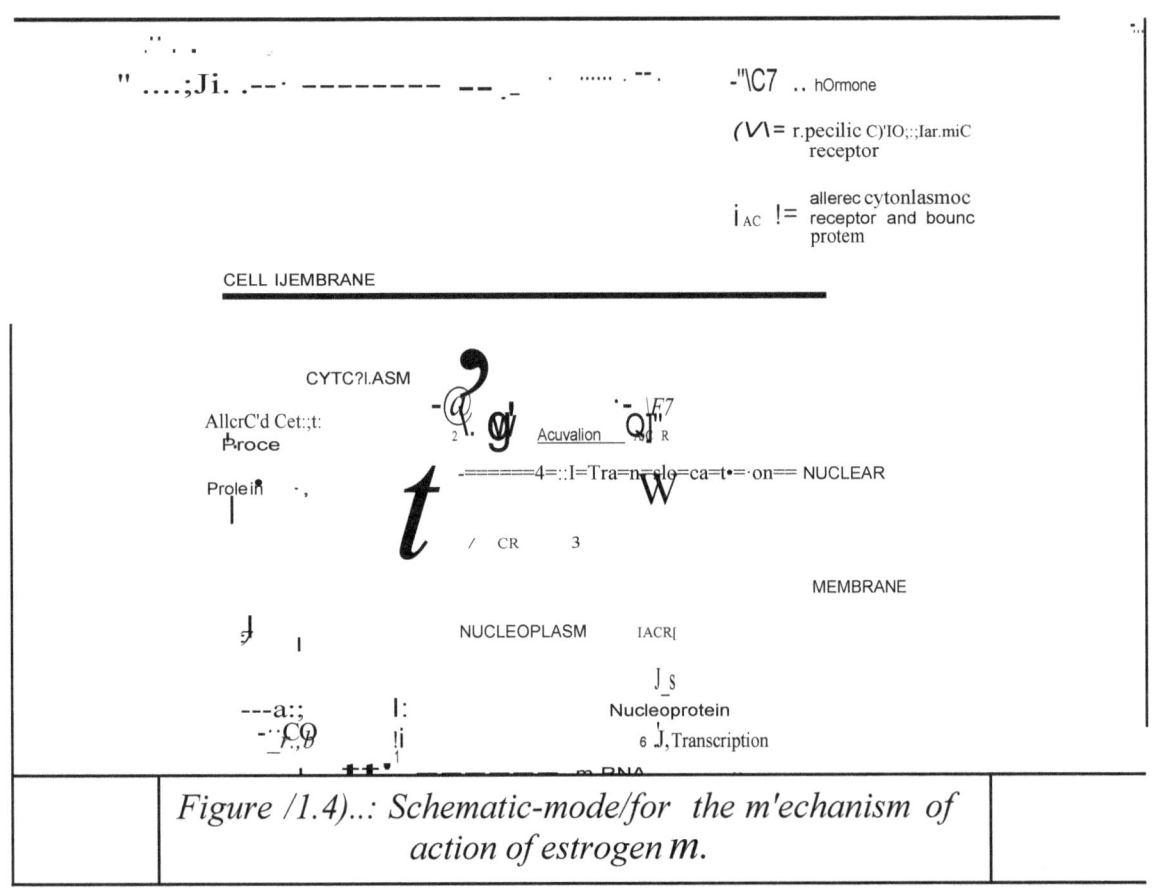

Figure /1.4)..: Schematic-mode/for the m'echanism of action of estrogen m.

The cellular response to increasing concentrations of steroid hormones is gradual and increases proportionally to hormone increase. Each receptor protein binds to a single molecule of hormone and each specific site on the chromatin binds to a single hormone-receptor complex. As the concentration of hormone increase, the concentration of hormone-receptor complex increases pr portionally, as does the number of complexes bound to specific sites on chromatin <S>.

1.1.1.7. Receptors

The estrogen receptor, also called estrophilin I has a molecular weight of about 200000 Da on binding of the estrogen molecule it undergoes a molecular change to yield estrophilin II, which may be regarded as the second messenger in estrog en action [16]>. The estro ens bind to the ligand site in the carboxy-terminal protein of the receptor molecule. This causes a conformational change that allows the receptor to bind to DNA. The DNA-binding domain ofthe ER recognizes the sequence AGGTCAnnnTG CCT (the estrogen response element, or ERE) [18, 19>.

The estrogen receptor contains 595 amino acids [•.4⁰']. There are two forms of estrogen receptor (a and ß), ERa and ER ß are the products of two distinct genes. The different forms ofER have selective effects in different tissues [118, 191].

The estrogen receptors in the human mammary gland decrease during the development and growth of breast cancer. Measurement of the amount of estrogen receptors in a small sample of mammary tissues is a diagnostic aid in recognizing the stage of the disease and for designing treatment. Once recognized, early breast cancer can sometimes be checked by changing the balance of androgens and estrogens in the body [06>].

1.1.2. Estrogen level change during female reproductive cvcle

Early in the cycle, when the levels of estrogens are relatively constant and low, the FSH levels are rising and high and the LH levels are low. These high levels of FSH stimulate release of estrogens, particul rly estradiol. By days 7 to

8, the estradiol level is nsmg at a rapid rate, reaching its first peak before ovulation $_O$ •46>.

Increasing levels of estradiol feed back to the pituitary gland vta a negative-feed back mechanism. resultinQ. in decreased secretion of FSH and increased secretion of LH. This results in a marked increase in LH secretion. known as the LH surge [1151]• Estradiol reaches a maximum on the day before the LH peak. During the mid-cycle there is a peak rise of LH. The estradiol level drops considerably and then rises again after ovulation II. -G. [1].

After ovulation LH and FSH both continue to decline during the luteal phase, reaching their lo\vest points in the cycle toward its end. The most distinctive and important feature of the luteal phase is a tenfold increase in progesterone [140]. -m. As the corpus luteum regresses, the levels of both estradiol and progesterone begin to diminish.. The removal of the inhibitory effect of these two compounds results in the increase ofFSH which stimulates the growth of a new crop of follicles in the ovary [111]•

1.2. Tumors of the ovan'

1.2.1. Classification

No other organ in the human body productes so many different tumors as the ovary [147]>. Ovarian rumors are categorized according to the world health or2_anization (WHO) classification, which is based on the site of orirrin of the tumor [1481], Table (1..1)[1491].

Epithelial ovarian tumor arising from the germinal epithelial of the ovary and serous epithelial ovarian tumors are the most common [150] • [53], \\'here mucinous tumors usually are of low malignant potential, and only 10%> of all mucinous ovarian tumors are malignant neoplasms [30] . [32]. Borderline ovarian tumors (also kno\\n as ovarian tumors of low malignant potential) are a subgroup of epithelial_ ovarian tumors [154] [55]>.

Table (1.1)- World healtlz organization (WHO) classification of ovarian (49)

tu.no,s			
Surface epithelium	Serous tumors	Sreous C.J'Stadenoma (beni!zn)	
		Broderline Sreous Cl'Stadenoma	
		Sreous cystadenocarcinoma (maliJmant)	
	Mucinous tumors	Mucinous c}•stadenoma (benig11)	
		Broderlilze Mucinous tunwr	
		Mucinous cystadenocarcinoma (mali 11ant)	
	Endometrioid carcinoma		
	Clear cell adenocarcinoma		
	Brenner tumor		
	Undifferentiated carcinoma		
Germ cell. tumors	Teratoma	Benign (mature, adult)	cystic teratoma
			solid teratoma
		Mali llalll (immature)	
		Mo11odermal	
	Dvs erminoma		
	Endoderma/ sinus tumor		
	Choriocarcinoma		
	Others	(embryonal carcinoma, polyembryoma, mixed germ cell tumors)	
Sex cord-stromal tumors	Granulosa-theca cell tumors	Granulose-tlleca cell tumors	
		Thecoma	
		Fibroma	
	Sertoli-levdig cell tumors		
	Gonadob/astoma		
Metastatic tumors			

Serous and mucmous borderline tumors are most often seen in women aged between 20 and 40 years, but other types of epithelial tumors of borderline malignant are rare [55 56]. Tumors of borderline malignancy is the term currently used in the (WHO) classification of oYarian tumors to describe group of tumors which are intermediate in both behavior and histological features betv,reen being rumors and those obviously malignant ([56]).

Germ cell tumors constitute the second most common group of ovarian malignancies, with the majority occurring in woman less than 40 years of age. And it represent 15% - *20o/o* of all oYarian tumors [6 :. 63]. Teratomas are tumors \\ith one or more of the three embryologic layers, ectoderm, mesoderm, and endoderm. Desgerminomas are the most frequent malignant germ cell tumors and the most frequent ovarian malignancy in young woman <s::n. Sex cord stromal rumors are derived from the sex cords and specialized stroma of the developing ovary. Cumulatively, it represent approximately 6% of ovarian neoplasms [25. 52]. Metastases to the ovary are common. especially ifthe primary cancer is in the breast or gastrointestinal rract, and it account for approximately 5% of all ovarian tumor [148 52].

1.2.2. Incidence

Tumors of the ovary are common forms of neoplasia in woman [49]. Ovarian cancer is the fourth most common cause cancer death in woman, and it is the leading cause of death from gynecologic malignancies [59 . 64]. There are numerous types of ovarian tumors, both benign and malignant. About 80% are benign, and these occur mostly in young woman between 20 and 45 years. The malignant tumors are more common in older woman between 40 and 65 years c[25. 49. 58. 65])

Epithelial ovanan can er accounts for 90% of all cases of ovarian cancer. The tumors are derived from the oelomir: pitheliltm, or mesothelium (JO. [61].[65]. [66]>, and it is uncommon in worl1an younger than 40 years, increases to a peak in woman aged 60 to 64 years, then decre ses c:ǒ.[61,1]. Serous tumors account for 20% - 50% of all ovarian neoplasms and 35%- 40% ·of ovarian cancers. About 70% of serous tumors are benign, 5% -10% have borderline malignant potential, and 20% - 25% are malignant. Se:1ous cystadenomas o cur most frequently in woman 30-50 years of age [1521]. And serous carcinoma$ occur in \\'Oman after 40 . years of age and grow in frequency with advancing age Hi.[521].

1.2.3. Etiology

Risk factors for ovanan cancer are much less clear than for other genital tumors [149]>. But several factors have been associated with an increased or decreased risk of the disease. Genetic, enviromental, and reproductive factors are associated with the risk for the development of ovarian cancer ,[59].[6],[671]. Reproductive risk factors include those associated with an incresed number of ovulations in a woman s lifetime. Menstrual factors have been incounsistently associated with a risk for the development of ovarian cancer, an early age of menarche and late menopause have been associated with twofold increase in the risk of developing ovarian cancer ([59.][68][1.]

Each pregnancy reduces the ovarian cancer risk by about 10%, and breast feeding and tubal ligation also appear to reduce the risk [169].[70]>. Oral contraceptive reduce the risk of ovarian cancer in patients with a familial history of cancer and the general popullation [162][66][68.][69.][71.721]. Estrogen replacement after menopause does not appear to increase the risk of ovarian cancer [169]1 Familial and genetic associations have been reported but are uncommon, cases of familial ovarian cancer may constituts 5%- 10% of total cases [160][731]

.. Hereditary llonpolypoc;is colon r;at1Ci r (!-I]\.TJ>CC) syndrome is also known ru: ':'lncer farnily syndrmi:e .vllyn h sy:t:ri.l'ome II,

·• Site-specific ov:1rian capr;P.r f.mily syndrome is now iinked to the BRCAI gen and thought to be a vati:mt ofthP. bre:'lst-0\'(llian cancer syndrome.

In woman \Vith no family histo1y of ovarian cancer, the lifetime risk is 1.6%, where as a woman whh one affected first-degree relative has 5% lifetime with two or mere effected first· degree relatives, the risk is 7% [1][8]. Approximately 3'?1> of women with two or more affected first-degree relatives will have aher ditary ovanan cancP.r syndrome with a lifetime risk of 40o/o [r:"Q']. Mutations of BRCA1 and BRCA1 genes are responsible for most cases of familial ovarian cancer [16][u.oi.][73][77]. Women with a BRCAI gene mutation have at least a 50% lifetime risk of ovarian cancP.r.

The P53 tumor suppressor gr.he is one of the most frequently mutated genes in h an ca t:P.r, h1duding approximately 50% of ovarian carcinomas [180][8]n. Few r\ ta exist to associated specific environmental factors ·with development of ovarian canc'!r. However, one environmental risk factor that has been consistently associatf!d with an increased risk for the disease is the exposure to talc powder on the perineal area [16][8]. ·>. Surgical interventions that decrease a women's exposure to environmental [82][84]agents have been associated with alower rate of development of the disease. Hysterectomy and tubal ligation both decrease the risk of ovarian cancer; however, the reason for this is uncertain but may be due to ablating a potential entry of environmental agents intothe peritoneal cavity [1591].

Staging of ovarian cancer is based on the e>..'tent and location of disease found at surgical exploration [150] [6]m.

The most widely used classification is that of the international federation of gynaecology and obstetrics (FIGO), Table (1.2) [1851].

The basis for staging is proYided by the initial laparotomy. In patients with an early stage of disease, any suspicious lesion should be biopsied. The peritoneal surface. the liver, the diaphragm and the retroperitoneal lymph nodes must be examined. Ascitic fluid should be sampled and peritoneal washings should be collected for cytological examination <[85].

Diagnosis of ovarian cancer in its early stage is very difficult, therefore most reported cases (60% - 80%) of ovarian cancer when first diagnosed are stage III and IV [173] [791] Ultimately, accurate staging is of utmost importance for the patients further therapy and for discussing prognosis [150] and increase survival rates of both patients with low-stage cancer and patients with advanced cancer [1861].

1.2.5. Clinical teatures

Because of the lack of any specific symptoms, ova ian tumors are often large by the time the doctor is consulted [1411]. The defmitive diagnosis of ovarian cancer is based on the results of surgical exploration and pathologic review [150]. .o\ny pelvic mass in women that is more than 1 year postmenopausal is suspicious for ovarian cancer [185] [871]. Only with prompt surgical exploration can ovarian cancer be diagnosed early [1501].

Chapter One   *Introduction and Literature*
Survey

The symptoms of epithelial ovartan cancer are often vague and nonsp[

have had months of gastrointestinal symptoms, including belching, early satiety,

abdominal fullness, or dyspepsia. Abdominal pain and pressure may have been
noted to be transient. As the disease progresses, symptoms become more
specific and constant, related to pain and pressure caused by an enlarging pelvic
abdominal mass or increased abdominal girth secondary to intraperitoneal tumor
or ascites [150]>. Physical finding include ascites and abdominal masses [187]>.

In 75%of pationts, malignancy has spread beyond the ovary at diagnosis,
and 60%have disease that has spread beyond the pelvis. 37% of patients had
abdominal discomfort or pain, 35% had abdominal swelling or masses, and
f!!   15%expEJriencedvaginal bleeding. Gastrointestinal symptoms occurred in 10%,
and urinary tract symptoms were present in 1.5'%, in two other large series the
main presenting symptoms were pain (57%), abdominal distention (51%), and
vaginal bleeding (25o/o) [160]. [88] - [91]>. Abnormal uterine bleeding was noted in
-34%ofpatients with ovarian tumors [13]>.

1.2.6. Diagnosis

Approximately 75%of women with ovanan cancer are diagnosed with
advanced disease after regional or distant metastases have become established
[2][5]. [1]. The overall 5-year survival' is approximately 17%with distant metastases,

36%with local spread, and 89%with early disease. Diagnosis is made by bringing together a number of simple investigations [175].

The first and most important technique to be used is clinical examination; this will allow the clinician to make a presumption of cancer on the basis of recognition of a hard pelvic abdominal mass. Cancers can be solid and irregular although mixed solid and cystic lesions are normal. A smooth single cystic swelling will tend to be benign; approximately 80% of cystic swelling of the

ovary in a premenopausal women will be benign [175]. While tumors with multiple

internal echoes or solid parts must be considered as malignant [157 92].

Ultrasound examination of the abdomen and pelvic has become a standard diagnostic test for evaluating pelvic masses. Ultrasound may determine the origin of the pelvic mass, delineate the internal appearance of the mass, and define associated abdominal finding, the presence of septations papillary structures, solid tissue. and bilaterality suggests malignancy in an o·varian mass. Furthermore, ascites and omental, nodal or hepatic metastases are strongly indicative of malignancy. However, the sensitivity of ultrasound in diagnosing ovarian cancer in premenopausal and postmenopausal women \\'ith adnexal masses was only 50% and 78%, respectively m.

Lumphangiography may be useful in evaluating patients with ovanan carcinoma and is positive in about 30% of patients studies. Lumphangiography is accurate when aonic lymph nodes are enlarged or replaced by tumor and radiologist has sufficient expertise [60 . 93. 941]. Computed tomography (CT) adds useful diagnostic and staging information to the results of ultrasonography, lumphangiography, and surgery [160. 951]. CT can delineate liver and pulmonary modules, large abdominal and pelvic masses and retroperitoneal nodal involvement. CT is coastly, however, and it cannot reliably detect masses smallerthan 2 em in diameter [1601] . .

Exploratory laprotomy has been the standard approach. Laparoscopy may be considered for a premenopausal women with an ovarian mass small enough to be removed using alaproscopic approach [196-991]. If malignancy is suspected, preoperative workup should include chest x-ray, evaluation of liver and kidney function, and hematologic indices c[981]. The technique of color doppler will play an increasingly importantrole in the differential diagnosis of a pelvic mass, and will serve to improve the specificity of ultrasound for ovarian cancer. It is yet to be shown that there use as a secondary test [000-1011].

In ovarian tumors, several reports have shown the usefulness of trans vaginal (TV)-CDU for differentiating malignancy from benign ovarian tumors [11011].

1.2.7. Tumor markers

Tumor markers are substances of different chemical nature that are either produced by a tumor or released froin cells in response to the presence of tumor, therefore disappearance of most of the markers should indicate eradication of the tumor [1102 - 104]. Tumor markers might be useful for the diagnosis of ovarian cancer and for monitoring the disease status during and after treatment. Quantitative and qualitative changes in numerous circulating substances have been associated with epithelial ovarian cancer. These may reflect an alteration in ovarian function or surface molecular structure, or a general response to malignancy [000]. Changes in circulating enzymes [1105 - 106] (galactosyltransferase. alkaline phosphatase placental, creatine kinase); hormones [11051] (HCG, estrogen, progesterone); antigens nos.wi-IIO) (CA l25, CEA. CA-19.9, CA-15.3, TAG-72) have all been identified in women with ovarian cancer.

Ovarian tumor- associated antigens have been reported, and experimental data suggest that they could be clinically useful in establishing initial diagnosis and m momtoring response to therapy [011 - 11..]. A part from the assoctatton of alpha- feto protein (AFP) with germ cell ovarian cancer (especially endodermal sinus and embryonal tumors) and human chorionic gonadotrophin with choriocarcinoma, these markers have not shown sufficient sensitivity or specificity for early epithelial ovarian cancer to be of value as screening test [000]. The most valuable tumor marker in the field of ovarian cancer is CA-125 [60 75]

CA-125 is a high molecular weight glycoprotein, which circulates in evaluated levels in the serum ofthe majority of patients with ovarian carcinoma, and is therefore of enormous value in the postoperative and chemotherapy phase of management [175 1131] This antigen is absent or very low in the serum of normal

· -125 appears to be elevated in approximately 80%ofpatients with non-mucinous epithelial ovarian cancer [150. 60 . 1 17]. Unfortunately, this

18

antigen is also elevated in patients with other malignancies and with such benign conditions as endometriosis, pregnancy, and pelvic inflammatory disease <50. [75]>. Nonetheless, a CA-125 assay may provide some guidance as to whether an adnexal mass is of benign or malignant nature. Studies to data suggest that more than 80%of menopausai women with an adnexal mass who also ha\·e an elevated CA-125 level are found to have ovarian malignancy. On the other hand, fewer than 10% of patients with a normal CA-125 level and an adnexal mass are found to have ovarian cancer [150]>. Serum CA-125 levels measured by radioimmunoassay are greater than 35 u/mL in over 80% of women with epithelial ovarian cancer (especially of the non-mucinous types) [000]>.

Carcinoembryonic antigen (CEA) levels are elevated in approximately 58%of patients with stage 1 1 1 epithelial ovarian cancer ([60. 118]>. The frequency of elevated CEA levels progressively increases with advancing stage and bulk of tumor. It is most likely to be elevated in patients with mucinous tumor [60]·. For patients with ovarian cancer who have elevated CEA levels before therapy serial CEA measurements may be valuable for monitoring subclinical persistence or relapse of disease [160. 119]>.

CA-125, CA15.3 and TAG·72 the most useful antigens for differentiating €17 benign from malignant disease. Serum level of all three antigens were ele\·ated in -77o/o of women with ovarian cancer compared with only 5%ofwomen with benign disease <[1]oo>.

1.2.8. Estrogen (Receptor) and Ovarian Tumor

The fact that the ovary is not only the main source of estrogens but also a target organ for these and other hormones suggests that epithelial ovarian cancer may be an endocrine- related tumor n.2o. [120]. Estrogen receptors are reported in approximately 50% of tumors. These lead to treat ovarian cancer patients with tamoxifen and progestins [1120. 122]. While, the majority (98%) of ovarian cancers contains androgen receptors. This predominance of androgen receptors compared to ERs may be due to the fact that the postmenopsusal ovary, which produces little or no estrogens, continues to produce androgens [122-125].

Cellular receptors for the estrogens have been shown to have therapeutic and prognostic significance in carcinomas of the ovary [126-129]. There are a correlation between the receptor concentration and the response to chemotherapy and the length of survival but no correlation with histopathology. grade, or stage [130].

A postmenopausal bleeding from an estrogen stimulatedendometriurn may, thus, be the first sign of an ovarian rumor. Interestingly, estrogen concentrations determined serially in serum from patients with ovarian cancer may reflect the progression or remission of rumor under treatment [031]. On patients undergoing surgery of ovarian rumors have shown that concentration of estradiol in venous blood from the ovarian tumor is higher than that of venous blood from the contralateral oYary and that of perpheral venous blood [131]. Previous studies have indicated that serum estradiol is good indicator of hormonal activity in ovarian tumor in postmenopausal women [130,132,133].

Epidemiological studies have yielded conflicting results about the risk of ovarian cancer with postmenopausal estrogen use. The mortality rate from ovanan cancer was about twice as high among woman who had taken postmenopausal hormone therapy (mainly unopposed estrogens) for 10 or more years compared with the rate among woman who had not taken hormone therapy. And no association with fewer than 10 years of use [134]. Tumors can also produce estroidal products. Estrogens from estrogen-producing tumors in the female cause amenorrhea, often in early symptom of an ovarian tumors. Besides amenorrhea, estrogen-producing tumors can cause anovulatory bleeding in pre menopausal v.·oman. irregular bleeding after menopause, and precocious puberty in children [15].

work

The aim of the work in this thesis includes the following:

1. Determination of estradiol level in sera of normal women and patients affected by serous ovarian tumors.

2. Molecular characterization of the binding of $\Box\Box^5$1-estradiol with receptors in benign and malignant ovarian tumors such as those of binding capacity and the. effect of various factors (receptor concentration, hormone concentration, pH, time, temperature, different salts and halides).

3. Determination of the kinetic and thermodynamic parameters of the binding reaction of estradiol with the receptors in ovarian tumors.

4. Isolation of estradiol receptors of ovarian tumor homogenates in patients of benign and malignant tumors.

5. Spectroscopic studies on the isolated estradiol receptors in patients with benign and malignant tumors.

Chapter Two
Experimental Work

2.1. Chemicals, instruments and samples

2.1.1. Chemicals

All common laboratory chemicals and reagents were of analar grqde and were used without further purification. Tris [hydroxymethyl] amino-methane. MgCh, MnCh, NaCl, Urea, PEG M.W (I 0000), NaBr, Charcoal, Bovine serum albumin, and Gelatin were obtained from Fluka.

EDTA (disodium salt), Na, K-tartarate, $CuSO_4 .5H_2O$, NaF, Nal, ZnCh, $CaCl_2$, Glycerol and f3-mercaptoethanol were obtained from BDH.

Sephadex G-150, Dextran T-70, Blue dextran .2000 were obtained from Pharmacia Fine Chemicals. Switzerland.

Radioimmunoassay Kit of radioactive estradiol e^{25}1- estradiol) was purchased from Immunotech international (FRANCE). The activity of labeled estradiol was approximately 8.5 J.LCi. The Kit is supplied with, one vial containing the labeled hormone (^{125}1- estradiol), seven vials of estradiol standards (unlabelled) ranging in concentrations from (0-5000) pglmL.

2.1.2. Instruments

The instruments used m this work were, LK.B gamma counter type 1270 Rack gamma II, cooling centrifuge type Hettich, Shimadzu U.V-Visible recorder spectrophotometer type U.V.-160, Pye-Unicam pH meter, Memmert water bath, SM-Shaker, and Memrnert incubator.

2.1.3. Patients

Two groups of ovarian tumors patients were included in this study. Group I contained 13 premenopausl patients with benign serous ovarian tumor, cystadenoma, (Age=27-35 years). Group II consisted of6 premenopausl patients with malignant serous ovarian tumor, cystadenocarcinoma, (Age=29-36 years), the groups were matched "ith a group control subjects (group Ill). All patients were admitted for treatment to (Baghdad Medical City, Baghdad Teaching Hospital), (Iraqi Medical College Hospital of AL-Nahrian University·). They were histologically pro,·en from the supervision of specialists Dr. Nawal Alash. Dr. Raji Al-Hadithi and Dr. Luay Edward. The patients were newly diagnosed and didn't undergo of any type oftherapy. Patients suffered from any disease like hypertension or diabetsthat may interfere with our study were excluded. The host information of all patients and normal healthy subjects is summarized in Table (2.1).

Table (2.1) - The host information of ovarian tumor patients and healthy sub"}ects stud"1 ed•

Grou(Js	Number	Age lvear)	Type of tumors
Benign	13	2i-35	• Benign ovarian tumor
Malignant	6	29-36	· 4 patients with ovarian adenocarcinoma stage *III*.
Control	16	24-36	

2.1.4. Preparation o(blood samples

Five milliliters of blood samples were obtained from patients underwent ovariactomy by venipuncture just before surgery. Sixteen physically normal age matched volunteers were used as controls. Blood samples were left for 20 min at

_room temperature. After coagulation, sera were separated by centrifugation at 2000 xg (4200 r. p. m) for 10 min then sera were aspirated and stored in capped

2.1.5. Collection o(ovarian tissue specimens

The tumors tissue were surgically removed from ovarian tumors by ovariactomy. The specimens were cut off and immediately rinsed with ice-cold isotonic saline solution. They were collected individualJy in plastic receptacle and stored at-20°C until homogenization.

•Preparation of uterine tumor tissue homogenate

The frozen tissues were weighed, sliced finally a scalped in Petri dish standing on ice bath, the slices were thawed and further minced with scissors then homogenized in TEMG buffer with a ratio of I :5 (weight: volume) by using a manual homogenizer. The homogenate was filtered through four layers of nylon gauze in order to eliminate fibers of connective tissues. Then centrifuged at 2000xg for 75 min at 4°C. The sediment was suspended in 10 volumes of TEMG buffer for 15 min at 4°C and then suspension was used to obtain the crude nuclear fraction and the supernatant was used as crude cytosolic fraction.

2.1.6. Buffers and reagents

All buffer solutions were prepared (IJSJ by dissolving the appropriate amount of salt in distilled water and the required pH was adjusted.

1. TEMG buffer (pH 7.4): 0.01 M Tris buffer containing 1.5 mM Na_2-EDTA, 2 mM f3-mercaptoethanol and 10% glycerol. The buffer was prepared by an appropriate dilution of the stock solution to 250 mL.

2. Dextran-coated charcoal (DCC) suspension.

This suspension was prepared by dissolving the following compounds: 1.25g charcoal, 0.625g dextran T-70 and 0.2g gelatin in 1OOmL ofTEMG buffer pH 7.4.

2.2. Determination of total protein content in serous ovarian tumors homogenates

The Total protein of ovarian tumors homogenate was determined by the method of Lowry et al '[136], using bovine serum albumin (BSA) as the standard protein. The details of the method are according to the following steps.

1. One milliliter of each standard bovine serum albumin (zero, 20, 40. 80, 120, 160. 200 J..Lg/ml) was pipetted in a set of duplicate tubes.

2. One milliliter of (1:10) diluted tumor homogenate was also pipetted m duplicate tubes.

3. Five milliliters of reagent C was added to all assay tubes.

4. The tubes were shaked and allowed to stand at room temperature for 10 min.

5. Half milliliter of reagent D was added to all assay tubes and mixed immediately.

6. The tubes were left at room temperature for 30min.

7. The absorbance of the blue solution was read at 750nm against an appropriate blank.

8.. The standard curve was obtained by plotting the absorbance against the orresponding concentration of standard protein as shovvn in Figure (2.1), and used ih the determination of the unknov.n protein concentration of the ovarian tumors homo2..enates.

Solution:

1. Reagent A: Alkaline sodium carbonate solution (2% Na_2CO_3 m 0.1 N NaOH).

2. Reagent B: Copper sulphate-sodium potassium tartarate solution (0.5% CuS04.5H20 in 1 lo Na. K-tartarate). This solution was prepared freshly by dissolving 0.1 g of Na, K-tartarate in 1OmL of 5% $CuSO_4.5H_2O$.

3. Reagent C: Alkaline copper solution, this reagent was prepared by mixing 50mL of reagent A with lmL of reagent B.

4. Reagent D: Falin Ciocalteau reagent was prepared by the dilution of the commercial reagent with an equal volume of distilled water on the day of use.

5.. Standard bovine serum albumin (stock BSA 0.2mglml): working BSA solutions were prepared by serial dilution of stock solution.

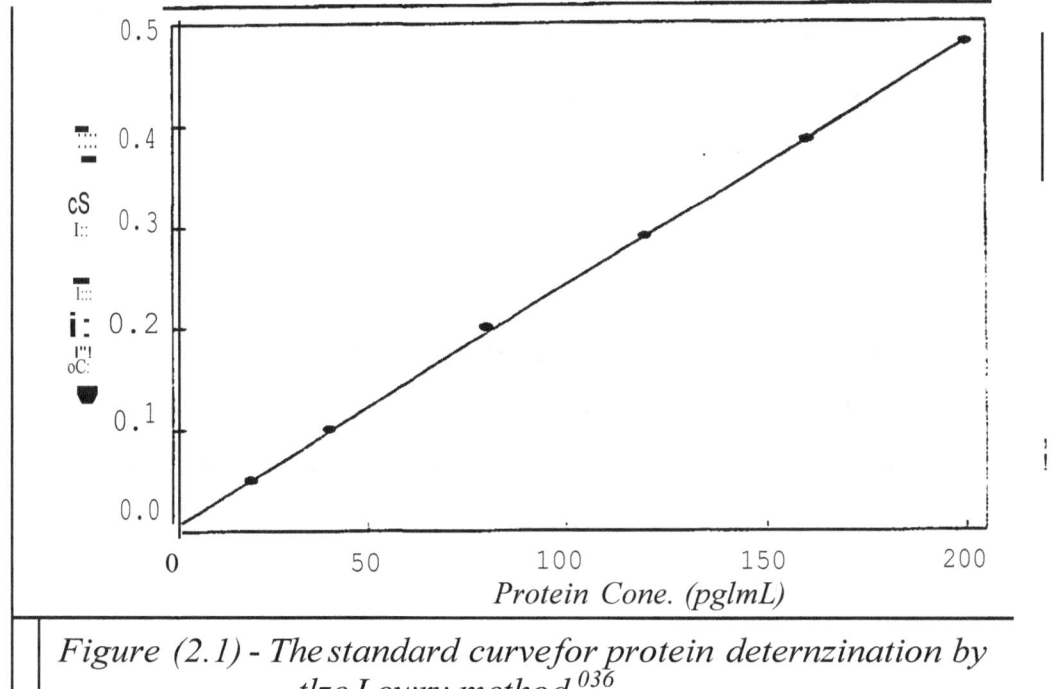

Figure (2.1) - The standard curve for protein deternzination by tlze Lowry method [036].

2.3. *Determination of estradiol levels in sera of benign and malignant serous ovarian tumors patients and controls*

Estradiol levels were measured in sera ofbenign and malignan o\·arian tumors patients and healthy individual by radioimmunoassay (RIA). The assay protocol was described in Table (2.2).

Table (2.2) -RIA assay protocol of serum estradiol (pglml).

---- ··-·-' - ··Estradiol standardS {;,si/;..:l)							Unknown			
	o	15	40 1150	5oo 11600 15000	(JJ	(2J	(3) : f4J
6Nited TUJieNo.	1.2	3.4	5,6 | 7,8 | 9.10	IJ,12 i13,1415,16	17,18	19,20	21.22
Standard	100	00	100 ! 100 ! 100 | 100 i 100	- | -	-	-
Sam_l!les	- | -	- | _ | _ ! - : -	100 | 100	100	100
-125I- estradiol.	500 : 500 1500: 500 : 500 | 500 500 ! 500	500	500	500
All volumes are in uL							

All tubes were then mixed gently after that incubated at room temperature (25°C) with gentle horizontal shaking (350 r.p.m) for 3hours. Then the tubes were aspirated and counted by y-counter for one minute<[137]), to prepare total count, SOOJ..LL of $^{1.5}1$- estradiol was pipetted in to aseparate tube and the radioactivity was determined by using y-counter.

Solution:

All solutions provided in estradiol RIA kit from immunotech international (FRANCE) were described previously in section (2.1.1).

Conclusions:

1. (B) is the bound radioactivity (CPM) which represents the counted radioactivity in the precipitated hormone-antibody compiex.

2. (B!f)% values were plotted against the concentration of standard unlabelled estradiol. Figure (2.2).

3. Results for the unknov.ns may then be read from the curve by intercept of the straight line as in Figure (2.2).

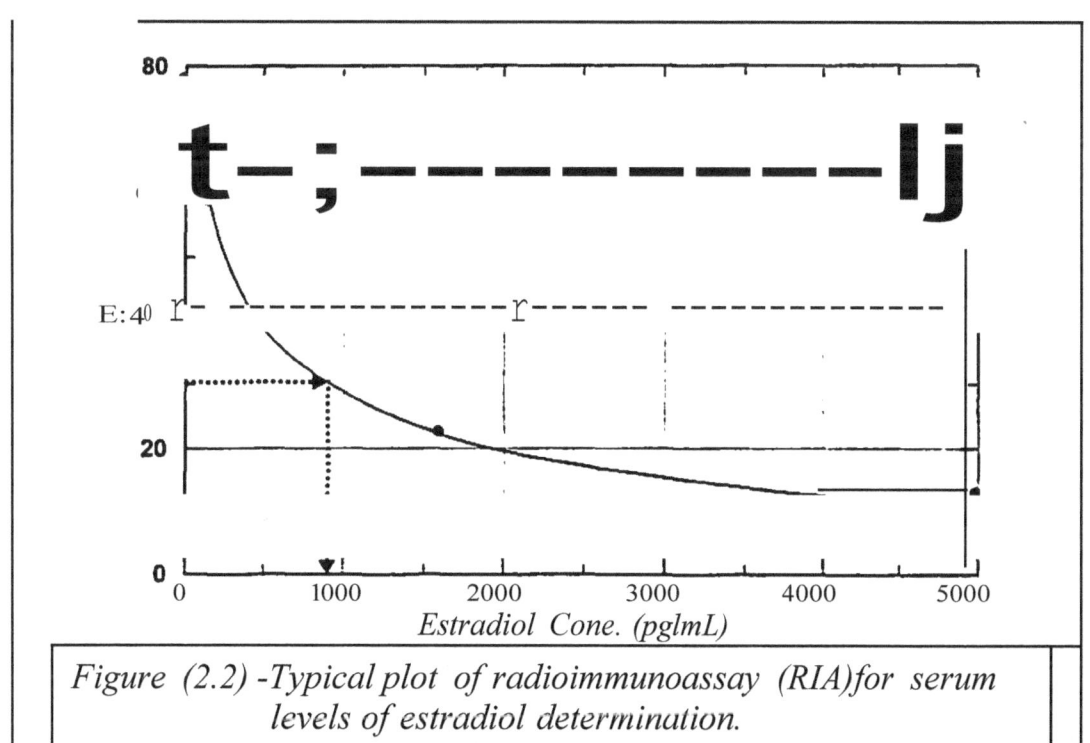

Figure (2.2) -Typical plot of radioimmunoassay (RIA)for serum levels of estradiol determination.

2.4. Determination of ^{125}I-estradiol concentration

The concentration of labeled estradiol was measured according to the method of Morris [038],\Vhich is summarized by the following steps:

1- In a set of coated tubes marked from 1 to 14. 500J,.LL of ^{125}I- estradiol was added with 100 J.LL of standard unlabeled estradiol of different concentration (0, 15, 40, 150,500,1600 and 5000 pg/mL). All tubes were then vortex gently.

2- In another set of the same tubes marked from 15-24, different volumes (250, 350. 450, 650, 850, and 11OOJ,.LL) of ^{125}I- estradiol were added with 100 J..LL of zero estradiol standard. All tubes were then vortex sz.entlv.

3- All the assay tubes were incubated at room temperature (25°C) with gentle horizontal shaking (350 r. p. m) for 3 hours.

4- All the assay tubes were aspirated and counted by ;-counter for one minute.

5- To prepare total counL different amounts of ^{125}I- estradiol (250. 350. 450, 650. 850, 1000) were pipetted in to aseparate set of tubes and the radioactivity was determined by usingy-counter.

Solution:

All solutions provided in estradiol RIA kit. from immunofech international (FR..c\..CE) were described pre\·iously in section c2.1.1).

Calculations

1. (B) is the bound radioactivity (CPM) which represents the counted radioactivity in the precipitated hormone-antibody complex.

2. The free hormone (F) which represents unbound ^{125}I- estradiol was determined from the followine formula:

F fCPM) =total count (CPM) -B (CPM)

3. The values of the ratio (B/F) for an ordinary standard curve were calculated. This standard curve represents the incubation of different amounts of estradiol standards with constant amount of ^{125}I-estradiol as in table (2.3).

4. The (B/F) values for the incubation of different amounts of ^{125}I- estradiol were also calculated. Table 2.4.

5. The data in tables (2.3) and (2.4) were plotted as in Figure (2.3) curve I and II.

6. The amount of radioactivity corresponding to the concentration of unlabled hormone was estimated by using the two curves (I and II) of Figure (2.3). This was carried out by drawing a line which intercepts with both curves at the same increment.

7. Table (2;5) shows the amount of standard estradiol in pg/ml corresponding to a given amount of tracer ^{125}I- estradiol.

8. The amount of the standard estradiol was plotted against the amount of corresponding radioactivity, Table (2.5) and Figure (2.4).

9. The concentration of ^{125}I- estradiol was determined from the intercept of the straight line as in Figure (2.4).

Table (2.3) - BIF values corresponding to different concentration of standard estradiol used in standard curve.

· ?6;,{i,1 ;;. a dr l;;;Ji+..M:	
0	3.15
15	3.063
40	2.845
150	1.17
500	0.65
1600	0.335
5000	0.187

Table (2.4) - BIF values corresponding to different amounts of ^{115}1- estradiol used in the incubation.

Amount ()(^{125}I- estradioi(JJL)	Bowzd radioactivitJr (CPMJ	Bff
250	3526	2.12
350	4287	1.93
450	5230	1.78
650	6515	1.04
850	7441	0.76
1100	8580	0.54

Table (2.5) - The amount of standard estradiol in pglml corresponding to tl given amount of the tracer ^{125}I- estradiol.

Bound radioactivity (cPMJ .	Hormone ·concentration foeimiJ
7109	250
7354	350
7696	500
7812	600
8148	700
8403	800

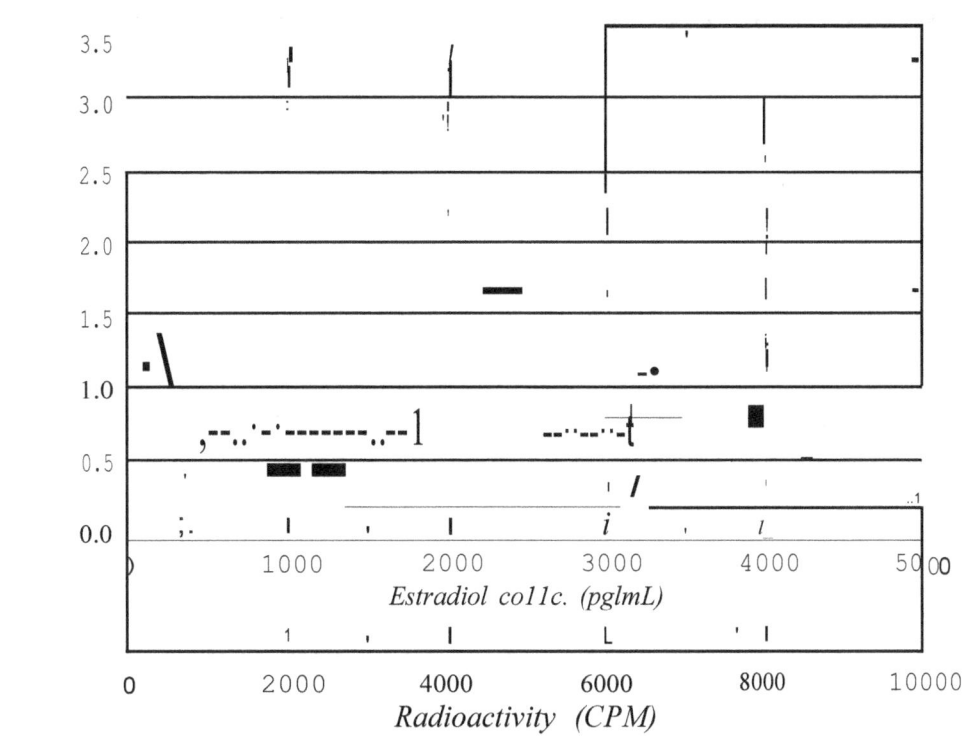

Figure (2.3) - Ratio of bound radioactivity to free radioactivity (BIF)
for an ordinary standard curve, where different
amount of estradiol standard were incubated with
constant amount of ^{115}/-estradiol and antibody (/), also
(BIF) for antibody incubated with different amounts oj
^{115}/-estradio/ in the absence of unlabeled hormone (///).

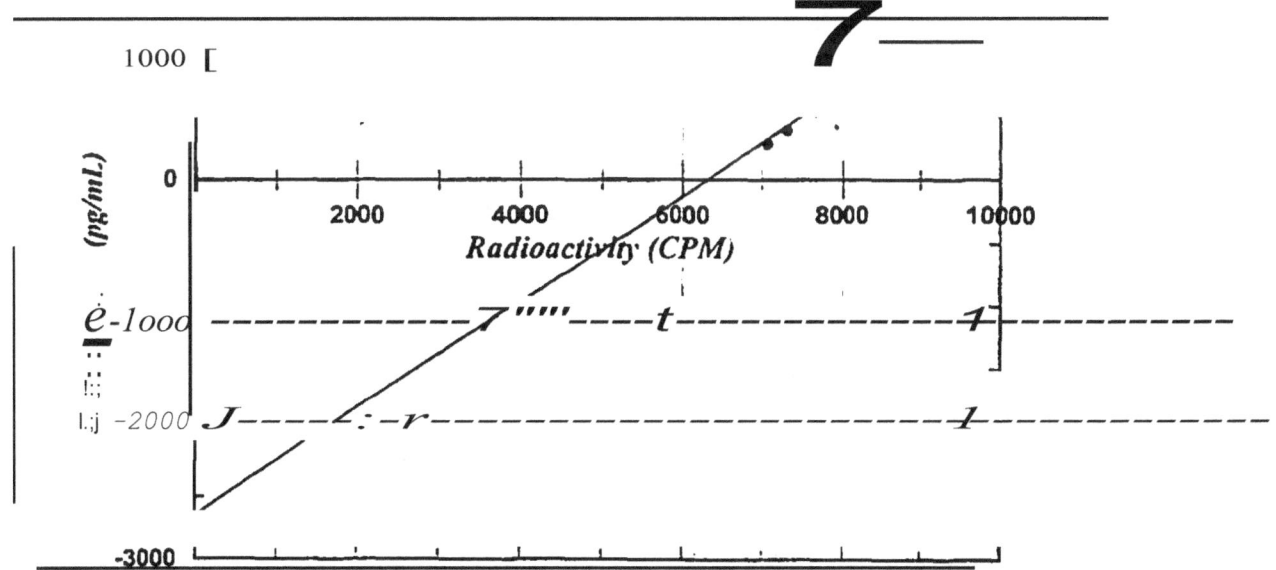

Figure (2.4) - A plot of the amount of standard estradiol against the
(CPM) of ^{115}/-estradiol having the same (BIF) from
Figure (2.3), which resulted a straight line. The
interest on the ordinate corresponds of the
concentration of ^{125}/-estradiol (pglmL).

2.5. Binding studies.o(^{125}1- estradiol witlz its receptors in benig11 and malignant serous ovarian tumors llomogenates

2.5.1. Preliminan' test of ^{125}1- estradiol binding to its cvtosolic and nuclear receptors in benign and malignant ovarian tumors

Cytosolic and nuclear receptors were evaluated using two sets of duplicate tubes. The first set was carried out to determine the total binding, while the second was used for the estimation of non-specific binding.

In order to detect cytosolic receptors l00J..1L *(250 J...Lg* protein of benign and malignant ovarian tumors) of crude cytosol were incubated with 50 J..1L (265pg/mL) of ^{125}1- estradiol in duplicate tubes. The volumes of the mixtures were completed to 500 J.lL with TEMG buffer pH 7.4 and the tubes were incubated for 16 hrs at 4°C. Non-specific binding was accounted by preparing the same incubation with the addition of 200 fold excess of unlabeled estradiol as a competitor. After incubation, the bound estradiol was estimated by the dextran-coated charcoal method [160]>. For this purpose 125J..1L of dextran-coated charcoal (DCC) were added. The ubes were shaken for 10 min. and centrifuged for 10 min. at 2000 xg (4200 r.p.m.) [1]II. An aliquot of500 J..lL was taken from each supernatant and counted by y-counter.

Crude nuclear receptors ere evaluated by the addition of l00J..1L (380J..1g protein of benign and malignant ovarian tumors) of crude nuclear fraction to 50 J..ll (265pglmL) of ^{125}1-estradiol with and without the addition of 200 fold excess of unlabeled estradiol. The final volume of the mixture was completed with TEMG buffer at pH 7.4 to 500 J..1L, the mixtures were incubated at 4°C for 16 hrs. During this time the tubes were vortexed several times. At the end of a period of incubation, bound and unbound estradiol were separated by charcoal adsorption with a suspension of dextran-coated charcoal. For these purpose 125J..1L of dextran-coated charcoal (DCC) were added. The tubes were shaken for 10 min and then centrifuged for 10 min at 2000 xg at 4°C. 500 J..LL of each supernatant was taken and counted by y-counter. It represents nuclear bound estradiol

Solutions

All solutions were prepared as described previously in section (2.1.6).

Calculations

1. The radioactivity in each tube (expressed in cpm) represents the total binding (TB).

2. The radioactivity (expressed in cpm) in the rubes contained $^{125}1$- estradiol and excess of unlabeled estradiol represents the non-specific binding (SB).

3. The specific binding (expressed in cpm) was calculated by subtracting the radioactivity (expressed in cpm) obtained in the presence of unlabeled estradiol from that produced in the absence of unlabeled estradiol.

SB (cpm) = TB (cpm) -1 VSB (cpm)

4. The percent of specific binding (SB%) can be calculated from the following formula:

$$SB\% = \frac{SB(cpm)}{Tc(cpm)} \times 100$$

Where: *Tc* total count ofthe $^{125}1$- estradiol (expressed in cpm) used in each tube.

2.5.2. fl.adioreceptor studies o{^{125}I- estradiol binding to its receptors in benign an4_ malignant serous ovarian tumors

All the following experiments were carried out ·with two sets of duplicate tubes. The first one was used to estimate the total binding and the second to estimate the non-specific binding.

2.5.2.1. The effect of different protein concentrations o(estradiol receptors on the binding o(^{125}I- estradiol in ovarian tumor ho1nogenate

Fifty microliters (265pg/mL) of ^{115}1- estradiol were added to 100 J.lL of increasing amounts (60, 120 180, 240, 300, 360, 420J.1g protien) of crude benign and malignant cytosolic fractions in a final volume of 500J.1L (completed with TEMG buffer pH 7.4) v.'ith and without the addition of 200 fold excess of unlabeled estradioL At the end of incubation (16 hrs) at 4°C, the bound est";1d:e-! was estimated by adding 125J.l.L of DCC, then the tubes were shaken for 10minute and centrifuged for 10 minute at 4°C at 2000 xg (4200r.p.m.). SOOJ.l.L of each supernatant was taken and counted by y-counter. It represents the bound estradiol.

Solutions

All solutions were prepared as described previously in section (2.1.6)

Calculations

1. The percent of specific binding (SB'!to) was calculated according to the formula mentioned in section (2.5.1).

2. The percent of specific binding (SB%) was plotted against the amount of protein receptors concentration included in each mixture.

2.5.2.2. The e(feg_of different concentrations o(^{125}I- estradiol on the pindi[lg l;Yith i s.rt!£eptors in benign and malignant o_varian _ tumor hol lzogenate

Increasing c<;>ncentrations (132.5-662.5pg/mL) of ^{125}1- estradiol was each added to 1OO!J.L (360, 240Jlg protein for benign and malignant o'\·arian tumor respectively) of crude cytosolic fraction in the first set of tubes ·with a fmal volume of 500 L (completed with TEMG buffer pH 7.4). The ond set of tubes consists of the same reactants plus *200* fold excess of unlabeled estradiol.

After incubation for 16 hrs at 4°C, 125 J.LL of DCC were added in order to estimate the bound estradiol then the tubes were shaken for 10 min and centrifuged for 10 min at 2000 xg at 4°C. SOO L was taken from each supernatant and counted by y-counter. It represents the bqund estradiol.

Solutions ,

All solutions were prepared as described previously in section *(*2.1.6).

Calculations

1. The (SB%) was estimated as mentioned in section (2.5.1).
2. The value of (SB%) was plotted against the concentration of ^{125}1- estradiol

2.5.2.3. The- effect o(di(ferent pH on 1:^{5}1-estradiol binding with its receptQrs in ovarian tumors homogenate

Crude cytosolic fractions (360, 240 g of protein in 1OOJ,J.L of benign and malignant ovarian tumor homogenate respectively) were added to 75J.1L (397.5 pg/mL) of ^{125}1-estradiol with and without the addition of 200 fo1d excess of unlabeled estradiol. The volumes of the mixtures were made up to 500 J.LL with TEMG buffer of different pHs ranging from 6.8 to 8.2. The tubes were

incubated   at 4°C for 16 hrs. After incubation, the bound estradiol was estimated as mentioned in section (2.5.1).

Solutions

All solutions were prepared as described previously in section (2.1.6).

Calculations

1.   The (SB%) was estimated as mentioned in section (2.5.1) at each pH.
2.   The   percent   of   specific binding   (SBo/o) was
plotted   against   their corresponding pH.

....).... *Tlze Choice of most appropriate incubation time of ^{125}/. estradiol binding to its receptors in ovarian tumor homogenate*

Seventy   five   microliters   (397.Spg1mL) of $^{1.51-}$ srradiol was added to 1 OO (360, 240J.lg protein for benign and malignant oYarian tumor respectively) of crude cytosolic fraction in a final volume of 500J.1L (completed with TEMG buffer pH 7.6 for benign and malignant tumors) with and without the addition of 200 fold excess of unlabeled estradiol. The tubes were incubated at 4°C. For different time intervals (2,4,6,8_10.16,18, 20 *22hrs)* at the end of the incubation the bound hormone was estimated as mentioned in section (:.5.1).

Solutions

All solutions were prepared as described previously in section (2.1.6).

Calculations

1.The percent of specific binding *(SB%)* was estimated as mentioned in section
 (2.5.1) at each incubation time.

2.The (SB%) was plotted against their corresponding time.

2.5.2.5. The effect o(temperature on the binding o[^{125}1- estradiol to its receptors in ovarian tumors lzomogente

Seventy five microliters (397.5pglmL) of c^{5}I-estradiol. were added to 1OOJ.LL (360, 240J,.lg protein for benign and malignant ovarian tumor respectively) of crude cytosolic fraction in a final ...·olume of SOOJ,l.L (completed ·with TEMG buffer pH 7.6 for benign and malignant tumors) with and without the addition of 200 fold excess of unlabeled estradiol. After incubation for 18hrs (benign tumors) and for *20hrs* (malignant tumors) at 4°C. the bound estradiol was estimated as mentioned in sections (2.5.1)

The experiment was performed at different temperatures (4. 25. 37 and 45°C).

Solutions

All solutions were prepared as described previously in section (2.1.6).

Calculations

1. The percent of specific binding (SB%} was estimated as mentioned in section *(2.5.1)* at each incubation time.·

2. The (SB%) were planed against the different temperatures of incubation.

2.5.2.6. The effect of different halides on tlze binding of ^{125}- estradiol to its receptors in benign and malignant ovarian tumors

Seventy five microliters (397.5pg/mL) of ^{1}:$_!^{5}$!-estradiol were added to 1OOJ..LL *(360,* 240J..Lg protein) of crude benign and malignant cytosolic fractions respectively in a fi.nal volume of 500 Jl.L (completed with TEMG buffer pH 7.6 for benign and malignant tumors containing 0.1M of each of the following halides: NaF, NaCl, NaB and NaiJ. with and without the additio of 200 fold excess of unlabeled estradiol. After incubation for 18 hrs at 37°C (benign tumors) and for 20 hrs at 25°C (malignant tumors), the bound eStradiol was estimated as mentioned in section (2.5.1). The control was used '";thout the addition of any halide.

Solutions

Halides solutions were prepared in concentration of O.IM in TEMG buffer pH 7.6, 1.0497 gm ofNaF in *250mL* ofTE:t\1G buffer, 1.461gm ofNaCl in 250mL of TEMG buffer. 2.572gm of NaBr in 250mL of TE 1G buffer

3.750I**D**J1 Nal in 250mL ofTEMG buffer.

Calculations

1. The (SB%) was estimated according to section (2.5.1) at each halide.
2. The (SB%) was plotted against each type of halide.

2.5.2.7. The effect of monovalent cations on the binding o(^{125}1-estradiol with its receptors in benign and malignantlz uman ovarian tumors

To evaluate the effect of monovalent cations on the estradiol bindin$! with its human ovarian receptors, the experiment was performed at optimum conditions (protein concentration, pR ^{125}1- estradiol concentration, temperature time) as mentioned in section (2.5.2.6) with one exception that the reaction mixtures were completed to 500 with TE!\1G buffer containing *25mM* of different monovalent salts (LiCl, **C1,** KCl and CsCl). The bound estradiol was estimated as mentioned in section (2.5.1). The control was used without the addition of any salt.

Solutions

Monovalent salts were prepared in 25mM in TEMG buffer pH 7.6, 0.3775gm of LiCI in 250 mL ofTEMG buffer, 0.3343 g of Cl in 250mL of TEMG buffer, 0.466gm ofKCl in .250mL ofTE!\.1G buffer, 1.0522gm ofCsCI in 250mL ofTEMG buffer.

Calculations

1. The percent of specific binding (SB%) was estimated as mentioned in section (2.5.1) for each salt.
2. The (SB%) was plotted against each salt type.

2.5.2.8. The effect of divalent cations on the binding o(^{125}I- estradiol with its receptors in ovarian tumor homogenate

The experiment was performed at optimum conditions of pH time, temperature, ^{125}1- estradiol and protein concentration as mentioned in section (2.6.2.6) with one exception that the reaction mixtures were completed to 500J.1.L with TEMG buffer containing 25mM of each of the following salts: MnCh.4H$_2$0, CuS04.SH20, CaCh, MgCh.6H20 and ZnCh

The bound hormone was estimated as mentioned in section (.2.5.1). The control was used without the addition of any salt.

Solutions

Divalent salts were prepared in 25mM in TEMG buffer pH 7.6, 1..2369 gm MnCh in 250 ml ofTEMG buffer, 1.5597grn CuS0$_4$ in 250 ml ofTE!\1G buffer, 0.6936gm CaCh in 250 mL of TEMG buffer, 1.2706gm MgCl: m 250mL ofTEMG buffer, and 0.8518gm ZnCh in 250mL ofTEMG buffer.

Calculations

1. The percent of specific binding (SB%) was estimated as mentioned in section (2.5.1) for each salt.

∴. The (SB%) was plotted against each salt type.

2.5.2.9. *The effect of different polvetltvlene glvcol concentration on tlte binding of ^{125}1- estradiol with its ovarian tumors receptors*

The experiment m section (2.5.2.6) was repeated but the Yolumes were completed to 500J..LL with TEMG buffer containing various percents of polyethylene glycol (PEG-I 0000) ranging from 2 to I 0%. The bound estradiol ·was estimated as mentioned in section (2.5.1) .The control was used without the addition of PEG.

Solutions

The stock solution of PEG-I 0000 (1Oo/o) was prepared by dissolving I 0 g of PEG in IOOmL ofTEMG buffer pH 7.6; 2, 4, 6 8 and 10% ofPEG solutions ·were prepared by an appropriate serial dilution of the stock solution.

Calculations

1. The percent of specific binding (SB%) was estimated as mentioned in section

   (2.5.1) at each PEG percent.

2. The (SB%) values were plotted against PEG percents.

2.5.2.10. *The effect of different urea concentrations on tlze binding of ^{125}I- estradiol to its receptors in benign and malignant ovarian tumors*

The experiment in section *(2.5.2.6)* was repeated but the volumes were completed to 500J!L with TEMG buffer containing urea ranging in their concentrations from 0.25M to 4M. The bound honnone was estimated as mentioned in s ction (2.5.1). The control in this experiment was used without the addition of urea.

Solutions

The stock urea solution (8M) was prepared by dissolving 24g in 50mL of TEMG buffer pH 7.6; 0.' 5, 0.5, 1, 2, and 4M urea solutions were prepared by an appropriate serial dilution ofthe stock solution.

Calculations

1. The percent of specific binding (SB%) was estimated as mentioned in section (2.5.1) at each urea concentration.
2. The '(SB%) were plotted against the corresponding molar concentrations of urea solutions.

2.5.2.11. *Competitive effect of different concentrations of unlabeled estradiol, progesterone and testosterone on the binding of ^{125}I- estradiol to its receptors in ovarian tumors*

The experiment was performed at optimum conditions (protein concentration, ^{125}I- estradiol concentration, pH, temperature, time) as mentioned in section (2.5.2.6) with one exception that it was performed with and without the addition of increasing concentrations (50-400 ng/mL) of unlabeled estradiol. The bound estradiol was measured as described in section (2.5.1). The experiment was repeated with increasing concentrations of unlabeled progesterone and testosterone.

Solutions

TEMG was prepared as described previously in section (2.1.6).

Ca/cu/atio11s

The percent of relative specific binding (RSB%) was estimated from the following formula:

$$RSB\% = \frac{Specific\ binding\ of\ ^{115}I\text{ -estradiol}\ in\ the\ prsence\ of\ a\ competitor}{Specific\ binding\ of\ ^{''''}I - estradw/\ in\ the\ absence\ o^{if}\ a\ competitor} \quad x \quad 100$$

2.5.3. The kinetic and tlzerJnodl·namic studies o(o[^{125}1- estradiol binding with its receptors in benign and nzalignant o1·arian

tU11lOTS

2.5.3.1. The tinze course o($^{1.5}$/- estradiol binding with its receotors in benign and malignant ovarian tumors

SeYenty five microliters (397.5pglmL) of ^{15}1- estradiol were added to 1OOJ.!L (360.240 Jlg protein for benign and maliQnant ovariar: rumor respectively) of crude cytosolic fraction. in a tina! volume of 500J.1L ı completed with TE 1G buffer pH 7.6) with and without the addition of *200* fok excess of unlabeled estradiol. Incubation was carried out for several time inte:!'"':a!s (6. 8. 1 0, 16. 18, 20 and *22* hrs). After each time interval. the bound hormone was estimated as described pre\·iously in section (2.5.1). The above experiment was performed at four temperatures (4. *25* *3* 7 and 45 C), for determination the time course of the association of ^{1} 1-estradiol with its receptors in benign and malignant ovarian tumors at different temperatures,

Solutions

All solutions were prepared as described previously in section (2.1.6).

Calculations

I. The Yalue of ^{125}1-estradiol bound specifically in (picomole of ^{125}1- estradiol per mg of protein) was calculated according to the following formula:

$$\begin{array}{l}\text{Tile val. of speciji ally]} \\ \text{bound} \bullet \text{l-estradwl} \\ \text{(pmole } \textit{I} \text{ mg protei11)}\end{array} = \frac{\left(\begin{array}{c}\text{Specifically boundj} \times \text{(Incubation volume in)} \\ m\,_1\,\text{estradiol in (PM))} \qquad \text{(Liter)}\end{array}\right)}{\text{mas of protem } mncu^b \text{ at i0nme}^d\text{lum}}$$

$$\begin{array}{l}\text{II} \\ \text{(Specijical\{r bound)} \\ \text{I} \,^{1''}\text{I}\,\text{-}\frac{estradiol}{(PM)}\end{array} = \frac{\text{(Total binding (cpm))- (1\on- speq(ic bmdmg (cpm)) x}}{\text{Totalcounts(cpm)}} \quad \begin{array}{c}\text{Total} \\ \text{concentrarionof} \\ 1\,\text{-}e^{\cdot '''}\text{tradl'o/} \\ mmcu^b dw\; n \\ medium\end{array}$$

$$\begin{array}{l}\text{(Tize percent of} \quad \text{J} \quad \left[\left(\text{Total binding}\left(\text{cpm}\right)\right)_1\text{-}\left(\text{J} \bullet \text{on- spec}^{if_1}_1\text{c bn}^{d*}\text{mg}\left(\text{cpm}\right)\right)\cdot\right] \\ \text{llspecijic bmdmg} = \text{Total coums (cpm)oif m} \textit{I} \text{- estradiol used ineaclz tube !x} \\ \text{(SB o)}\end{array} \cdot 100$$

2. The plot of the values of SB% or the \'alues of ^{125}1- estradiol specifically bound in (PM) against the time intervals yielded the time course curYe for the association of the ^{125}1-estradiol with its receptors in ovarian tumors

2.5.3.2. Determination of estradiol receptors concentration and the a(finitv constant of ^{125}1-estradio/ association with its receptors in benign and 1nalignant ovarian tumors.

Crude cytosolic estradiol . receptors v.·ere measured according to the following: lncreasing concentrations (132.5-397.5pg/mL) of ^{125}1- estradiol was each added to 1OOJ.LL (360.240 J..Lg protein for benign and malignant ovarian tumor respectively) of crude cytosolic fraction, with and \Vithout the addition of 200 fold excess of unlabeled estradiol in a final volume of 500 J..LL (completed with TEMG buffer pH 7.6). The tubes were incubated for 18hrs at 37°C (benign tumors) and for 20hrs at 25°C (malignant tumors) in order to anain an equilibrium state. The bound estradiol was estimated as mentioned in section (2.5.I). The previous steps were performed at different temperatures (4. 25, 37 and 45°C). The time of incubation needed to get the equilibrium state were

18hrs at (25, 37and 45) and 20hrs at 4°C (benign tumors) and for 18hrs at l4, 37and 45) and 20hrs at 25°C⁰ (malignant tumors)

Solutions

All solutions were prepared as described pre\iously in section (.1.6).

Calculations

1. The concentration of receptors and the affinity constant *v.rere* determined accord· g to $S_{catch}ar^d$ equanon ᶜ ṇᶜ . l ṇ,:

$$\frac{B}{F} = \frac{1}{k_d}(B_{max} - B)$$

$$k_\varrho = \frac{1}{k_d}$$

\\'here:

B: The concentration of bound hormone specirically.

F: The concentration of free hormone.

Bmox: The maximal binding capacity.

Ka: The affinity constant.

Kd: The dissociation constant.

2.   The free   hormone   *(F J*   which   ᴾ⁻₋₌₋₁₋ estradiol. was represents unbound determined from the following formula:

F(CPM) =totalcount (CPM)-B (CPM)

3. The values of ¹²⁵l- estradiol which is bound specifically in picomolar were calculated usimr the followin_ formula:

$$B = \frac{Total\,binding - \Lambda \cdot on\text{-}specific\,binding}{Total\,count} \times concentration\,of\,1\text{-}estra^{d''}\,{}^{if}_{10}\,{}^{1''}{}_{1Y.1}\,I(P'liA)$$

in each assay tube

4. The Yalues of the ratio B'F were plotted against the values of the (B). the receptor concentration and the affinity constant „·ere calculated from the x-a.xis and the slope of the straight iine respecti,·ely.

2.5.3.3. The therlnodrnanzic studies o(o(^{115}/- estradiol binding -..:ith its receptors in benign and lna/ignant ovarian tlunors.

According to the steps of the experiment explained in section (2.5.3.: 'the thermodynamic parameters were calculated.

Calculations

1.   The thermodynamic parameters of standard state were obtained from \'an ·t Hoff plot. the Yalues of the natural logarithm of equilibrium constant ı af::..1iry constant   ka) obtained at differem temperatures were planed agains: the

reciprocal values of absolute temperature in Kel\·in (1/T)   according ıc the fol pwıng equation II-IIIı:

$$\ln k_{II} = \frac{..\Lambda SO}{R} - \frac{.1\}[0}{RT}$$

\\'here:*Lllr:* The enthalpy change of the standard state.

 :The entropy change of the standard state.

R: The gas constant (8.3144 J.mol"^{1}.K·^{1}).

• .1.W' \·alue was obtained from the slope of the linear relationship of the plo:.

• (LlG°J The change in Gibbs free energy ofthe standard state, was calculated from the following equation:

_jG" =-RTink,

- (6S⁰) the standard state entropy change, was calculated from the following formula:

$$\Delta S^o = \frac{\left(\Delta H^o - \Delta G^o\right)}{T}$$

2. The thermodynamic parameters of the transition state were optained from Arrhenius plot of ln k-₁ ,·alues against (1 T) Yalues, that giYes a linear relationship according to the following equation:

$$ln k_{-1} = lllA-(:J$$

\\·here:

A: The .1\.rrhenius constant.

Ea: The activation energy.

R: The gas constant.

T: Absolute temperature.

- The Yalue of apparent energy of acti\'ation (Ea) of the binding rea=:tion

    can be determined from the slope ofthe straight line.

- The enthalpy change of the transition state (.M{) was calculated from

    the following equation:

$$\Delta H^* = E_a - RT$$

- The transition state of free energy change (G·) was calculated from the

    following equation:

$$LtG^* = -RT \ln k_{+1} + RT \ln \left(kTJ\right.$$

\\'here:

k: Boltzmann constant$\{1.38 \times 10^{-13}$ J.deg"^{1}).

h: Plank constant $(0.66 \times 10^{-33}$ J.sec"\

- The transition state entropy change (..IS.) was calculated from the roliO\\·ing formula:

$$\Delta S^{*} = \frac{\left(\Delta H^{*} - \Delta G^{*}\right)}{T}$$

2.6. Isolation o[estradiol receptors hr gel filtration technique.

- *Preparation o(rhe column*

    The dimensions of the column were chosen according to the foliowing equations [11411]:

- *Dia1neter (em)* = $\frac{\cdot\ m}{\setminus\ 10}$.

\\'here:

n1 : the amount of protein in mg:

Length (em)= 30 x diameter (em)

    In view of the results of such calculation, a 0.7 x *23cm* coiu:t7lil has been used.

- *Gel preparation and column packing*

    The gel (sephadex G-150) was allowed to swell in excess *ofTEMG* buffer pH 7.6 containing 0.02% sodium azide (I g of gel in 50mL buffer) and left to stand for three days at room temperature without stirring to equilibrate v.rith the buffer. The gel slurry was degassed by suction, then the swolien gel was poured carefully into a venical glass-column down the wall using a glass-rod. After the gel has settled the column was equilibrated with TEMG buffer pH *1.6* containing 0.02% sodium azide for 24 hrs with the dimension of (0.7 x .,...,em)

- *Void l'olume (Vo) determination*

The void volume of the column was determined bv using blue dextran 2000 at concentration of (2 mg/ml) dissolved in TEMG buffer pH 7.6, 1 mL of blue dextran solution was applied with the same buffer, using a flo,,· rate of 6mL!hour. Fractions of 1 mL then collected. and their absorbances weremeasured at 600nm. to determine the void volume (V_o=7mL).

- *The preparation o(crtosolic salt ext*

The frozen tissues were weighed, pulverized finely with a scalpel in Petri dish standing on ice bath. and then homogenized at 4°(in TEMG buffer solution with a ratio of 1 :5 (weight: volume) using a manual homogenizer. The homogenate v.:as filtered through four lavers of rvlon gauze to remo\·e tissue clumps and fibers of connective tissues. Then centrifugation at 000 xg for 75min. The pellet was decanted. and the supernatant was used as crude cytosolic fraction.

- *Isolation procedure_*

One milliliter of the crude cytosolic extract (3.5mg protein) was applied to column equilibrated with TEMG buffer. The sample was eluted using the same buffer, fractions of 1mL were collected at a flow rate of 6 mL/hr. The absorbances of the fractions collected were measured at 280nm and the protein contents were determined by the method of Lowry et al [036]>.

• *Tlte preliminan· test of the binding of [15]1- estradiol to the isolated fractions separated by gel filtration*

A volume of **1OO** of isolated fractions were added to 75J.Ll (397.5 pg/mL) of [1]:"1- estradiol in a final Yolume of 500 L (completed with TE 1G buffer pH 7.6). \\7ith and without the addition of 200 fold excess of unlabeled estradiol. The tubes were incubated for 18 hrs at 37°C (benign tumor), and for **20** hrs at 25°C (malignant tumor). The bound estradiol was estmated as mentioned in section (:.5.1).

• *Diall·sis for concentration*

The fractions that contained high leYels of estradiol receptors \-;.·ere pooled and concenrrated by dialyzing against sucrose at 4°C for 30 min to get the needed concentration.

Calculations

1. The SBo/o for the eluted fractions were estimated as mentioned in se,:-:ion 12.5.1).

2. The \·alues of SB⁰·o and absorbances at 280nm were ploned agam . the fraction number.

3. The isolation fold for each estradiol receptor for benign and malig:oam ovarian tumors was estimated from the following formula:

$$1solatton\ fi_old = \frac{Specific\ binding\ of\ isolated\ receptor\ (fmole\ I\ mg\ prot.)}{Specific\ binding\ of\ crude\ receptor\ (fmole\ I\ mg\ prot.)}$$

2.7. Determination of the molecular weight of isolated estradiol receptors.

A sephadex G-150 column *(0.7x23cm)* was used for this purpose as described in section *(2.6).* One milliliter of standard protein solutions (fe tin

(440KD), catalase (232KD), aldolase (158KD), and BSA (67KD)) were applied separately onto the surface of sephadex G-150 column. The elution was carried out with TEMGbuffer pH7.6 at aflow rate of 6mL/hour. The absorbances of the fractions collected were measured at 280nm to e\"aluate the elution \·olume tYe *i* of the standard proteins. Cytosolic estradiol receptors were applied onto the calibrated sephadex G-150 column and eluted as described in section (2.6).

Calculations

1- The (Kav) values ofthe protiens eluted were determined using the following formula:

$$K_{0\cdot} = \frac{ve-v}{V_l -V}$$

\\'here:

$V\cdot$: void volume

J "e:elution volume.

V_l: the volume ofthe bed gel.

2- The values ofKa\· (for each standered protein)\vere plotted against the values of log !v1.wt ofthe protein eluted.

The molecular \\'eight of the estradiol receptors were calculated from the standard curve obtained.

2.8. *Spectroscopic studies of different isolated (orms of estradiol receptors.*

2.8.1. *Tlze U.V.spectra o(isolated estradiol receptors in benign and malignant ovarian tumors.*

One hundred microliters *(350J.lg* protein) of each isolated receptor was completed to 1mL with distilled water pH 7.4, then placed in a 1em cuvette in

sample beam and the absorption spectrum was immediately measured against the adjusted pH 7.4 distilled water as a reference.

2.8.2. *Factors affecting tlze absorption properties of isolated estradiol receptors in benign and 1nalignant ovarian tumors.*

2.8.2.1. *pH effect.*

One hundred microliters (350 g protein) of isolated receptors were completed to 1 ml with distilled water at different pH (2.7, 7.4 and 10.7) then each of which was placed in the test cell and the adjusted pH distilled water was placed in the reference cell and the absorption spectra of different isolated receptors were measured immediately.

2.8.2.2. *Polaritv effect*

+   *The effect of 20% ethanol on the estradiol receptors spectra:*

One hundred microliters (350 g protein) of isolated receptors were completed to 1mL with distilled water contain *20%* ethanol at pH 7.4 then each of which was placed in the test cell and the *20%* ethanol adjusted pH was placed in the reference cel1 using I em cu\'ene. The absorption spectrum of each sample was measured immediately.

+   *The effect of 20% ethylene glycol on the estradiol receptors spectra:*

One hundred microliters (350 g protein) of isolated receptors were completed to 1 ml with distilled water contains *20%* ethylene glycol at pH 7.4 then each f which was placed in the test cell and the 20% ethylene glycol adjusted pH was placed in the reference cell using 1em cuvette. The absorption spectrum of each sample was measured immediately.

+ *The effect of 20% urea on the estradiol receptors spectra:*

One hundred microliters (350)..lg protein) of isolated receptors \\'ere completed to 1 ml with distilled water at pH 7.4 containing 20'% urea then placed in the test cell against t.l-te .20o/o urea adjusted pH in the reference cell using 1em cuvette. The absorption spectra of different isolate receptors were measured immediately.

+ *The effect of 20% DMSO on the estradiol receptors spectra:*

One hundred microliters (350)..lg protein) of isolated receptors were completed to 1 ml with distilled water at pH 7A containing *20o/o* DMSO then placed in the test cell against the .20% DMSO adjusted pH in the reference cell using 1em cuvette. The absorption spectra of different isolate receptors were measured immediately.

2.8.3. *Spectrophoto1 1zetric pH titration o[isolated estradiol receptors in benign and malignant ovarian tumors*

A series of isolated receptors were completed to 1 ml with distilled water at pH range from 4 to 8.0. The maximum absorbance of each sample was measured at a wavelength of21lnm. The absorbance ofA.ma" at each pH value was plotted against the c rresponding pH.

Another series of isolated receptors (350 g protein m 1OOJ..ll) were completed to 1 ml with distilled water at pH ranging from 9.0 to 12.5. The maximum absorbance of each sample was measured at a wavelength of 295nm the absorbance of) .rna... at each pH value was plotted versus the corresponding pH.

2.8.4. *Tlte U.V.spectra o(di((erent estradiol -receptors co1nplexes of benign and malignant ovarian tumors.*

The binding experiment of different isolated estradiol receptors with t:sy_ esrradiol was carried out at the optimum conditions as explained previously in section *(2.5.2.6)*. One milliliter of the !-estradiol-receptor complex supernatant of each type of isolated receptors was placed in 1em cuvette in the sample beam and the absorption spectrum was measured immediately against an appropriate blank in the reference beam.

2.8.5. *The U.V-spectrum o(^{125}/- estradioL*

One milliliter of ^{1}::1-estradiol was placed in a 1em cuvene in the sample beam and the absorption spectrum was measured immediately against an appropriate blank in the reference beam.

2.9. *Statistical analvsis*

The results of serum estradiol was analyzed statistically and \'alues were expressed as mean $=$ SD. The levels of significance were determined (ı٧ student"s t-test.

Chapter Three
Results and Discussions

3.1. Preparation o(ovarian tumors tissues honzogenate.

Two groups of pre menopausal patients were included in this study. Group I contained (13) patients (luteal phase) with benign serous ovarian tumor. age !27-35). Group II contained (6) patients (luteal phase) with malignant serous O'vanan rumor, age (29-36) as confirmed by histopathological exammanon. Homogenization was carried out in cold medium in order to decrease the probability ofthe protein denaturation and the proteoiytic enzyme acti\·ity [11,1].

The homoQenate was filtered throuQh three nd on *Qauze* in order to eliminate any suspended pieces of unhomogenized tissue. fibers of connective nssue, and blood \'essels, \\·hile homogenate centrifugation at 2000xg (4200r.p.m.) precipitates the unruptured cells (the remaining intact cells ı and the intact nuclei of the ruptured cells" (crude nuclear fraction), leaYing other cytoplasmic constituents in the ·supernatant. (crude cytosolic fraction). The amount of protein in cured cytosolic fraction was 3.5mg mL for benign o\·arian rumor and *4.2mg/mL* for malignant ovarian tumor" as determined by Lo'"TY et al method [1136].

3.2. Determination of Estradiol Levels in Sera of Ovarian Tumors Patients

Estradiol levels m sera of patients with benign o\·arian tumors (group I) and malignant o\·arian tumors (group II) were measured by radioimmunoassay. The two groups were matched with a group of control subjects.

Table (3.1} and Figure (3.1) shows the results that were obtained from this study. The mean level of serum estradiol in patients ·with benign ovarian tumors \Vas found to be (258.7±38.9pg/ml), whereas that of control was found to be (212.3±30.4pg/ml). But in patients with malignant ovarian tumors, the mean le\·el was found to be (268.4+46.1pg/m1). Significant increase of serum estradiol levels in benign and malignant tumor ($p<0.005$) was obtained from Student's t-test analysis

Table (3.1) - Serunz estradiol lel•els (pglmL) in patients with benign ovarian tu11zors and nzalignant ovarian tunwrs. Details are described in section (2.2.1).

Group	No. of cases	A!!e (wuzr)	Serum estradiol (oelml)
I (benign ovarian tumors)	13	2":-35	258.7+ 38.9
II (malignant ovarian tumors)	6	29-36	268.4 + 46.1
Control	16	24-36	212.3 + 30.4

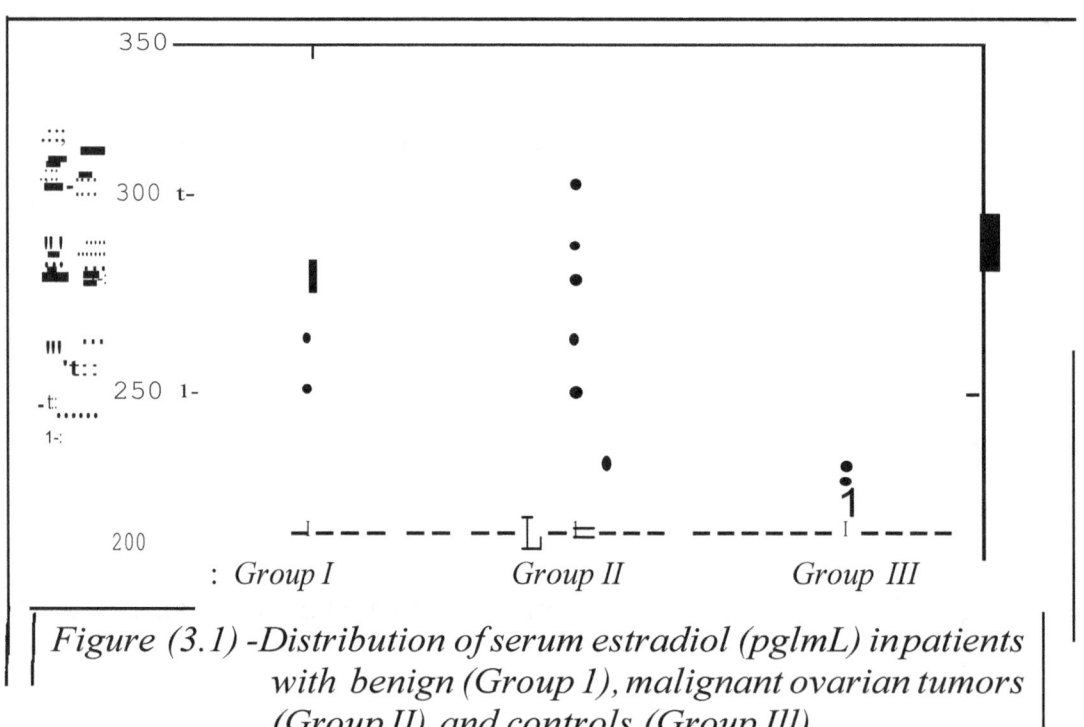

Figure (3.1) -Distribution of serum estradiol (pglmL) in patients with benign (Group 1), malignant ovarian tumors (Group II) and controls (Group Ill).

These results are nearly similar to those obtained previously by other investigators (JO. [134]\ Heinonen et. al, reported that there were no significant differences between the mean levels of estradiol in patients with benign or malignant tumors, but the mean serum estradiol levels in patients with benign or malignant ovarian tumors were hi her than those in the controls [11441]

3.3. Binding studies o(^{125}/-estradiol with its receptors in benign and malignant ovarian tu1nors homogenate.

3.3.1. Preliminarv test o(^{125}/-estradio/ binding to its cvtosolic and nuclear receptors in benign and nzalignant ovarian tumors homogenates

Cytosolic and nuclear receptors were investigated in benign and malignant oYarian tumors. Cytosolic estradiol receptors were detected through the incubation of ^{125}!-estradiol with crude cytosol fraction and the bound estradiol was separated by dextran-coated charcoal method' [127 . 1451]. and then measured by y-counter. Nuclear estradiol receptors were eYaluated using a nuclear fraction that incubated with labeled estradiol. The percent of specific binding in cytosolic

fraction was found to be 3.9% in benif!n ovarian tumors and 6.:!% in malim1ant ovarian tumors. 'A'hile there were no detectable levels of nuclear estradiol receptors in benign and malignant ovarian tumors.

Also it should be mentioned that there were many studies done on the ovaries show the binding of estradiol with its receptors on cytosolic ovarian tumor homogenate the data obtained in this study revealed that tumors of ovarian cancer patients have higher incidence of estradiol receptor than those of benign group, these result are consictent with those reponed previously. by Bergqvist A., Kullanders, and Thorell J.n [301].

3.3.2. Radioreceptor studies o(^{115}/-estradiol binding to its receptors in benign and malignant ovarian tumors homogenate

3.3.2.1. Tlte effect of different protein concentrations of estradiol receptors $_{011}$ the binding of ^{125}/-estradiol in ovarian tumor lzonzogenate

To determine whether the different protein concentrations of benign and malignant tumor homogenate affect the binding, increasing amounts of cytosolic homogenate were incubated with fixed amount of ^{1}^{5}1-estradiol with and without the addition of nonradioactive estradiol. according to the details in section (2.5.2.1). As shown in Figure (3.2). the specific binding percent was increased ·when the amount of receptor protein in the incubation mixture was increased to reach the maximum binding. These results indicate that estradiol receptors binding are principally depended on the amount of receptor protein in the reaction mixture In all the subsequent experiments. 360 g and 240 g of receptor protein of benign and malignant :rude cytosolic homogenate. respectively, were used.

To determine the maximum binding ability of ^{1}^{5}Ł-estradiol with its receptors in benign and malignant OYarian tumors homogenate. Line WeaYer-Burk relationship [11 61] was plotted (the data in Figure *(3.1)* was plotted as 1/SBO;o Vs l/J.lg protein). The y-intercept of the plot represent the in\·erse of the maximum amount of ^{125}1-estradiol that could be specifically bound to infinity receptor concentration, Figure (3.3).

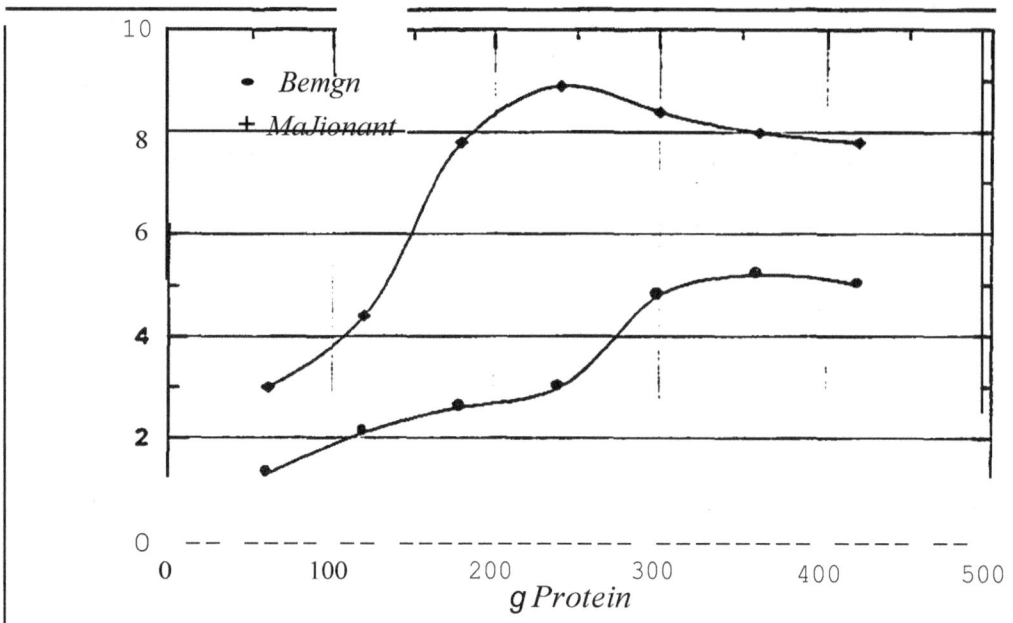

Figure (3.2) - Effect of different protein concentrations on the binding of u^5 /- estradiol to its receptors in ovarian tunzor homogenate.

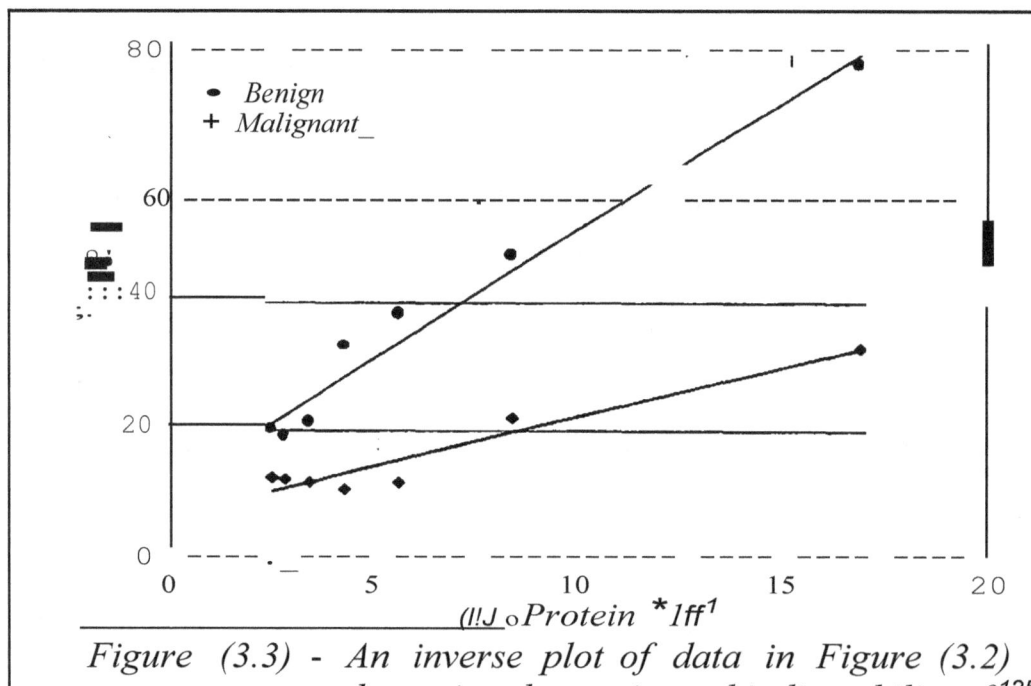

Figure (3.3) - An inverse plot of data in Figure (3.2) to determine the maximum binding ability of [125]J-estradiol with its receptors in benign and malignant ovarian tumors homogenate

3.3.2.2. *Tlze effect of different concentrations o(^{125}/-estradiol on the binding with its receptors in benign and malignant ovarian tumors*

One of the most important criteria of the true receptors is the sarurability. To fulfil this criterion and to estimate the suitable concentration of^{125}1-estradiol, th experiment was carried out as mentioned previously in section (2.5.2.2). The results are illustrated in Figure (3.4). It is shown that the specific binding of tracer estradiol with receptor   protein is a saturable process but complete saturation however is theoretically never reached unless the amount of estradiol used reached infmity o.r;-,. As sho·wn in the   same figure. the   receptor   was saturated with ^{125}1-estradiol molecules   at the   concentration   of 397.5pg mL (994 pM). So in the subsequent experiments 397.5p lmL of ^{125}1-estradiol was used in   the   incubation   mixture   for   both   benign   and   malignant   ovarian   tumors homogenate.

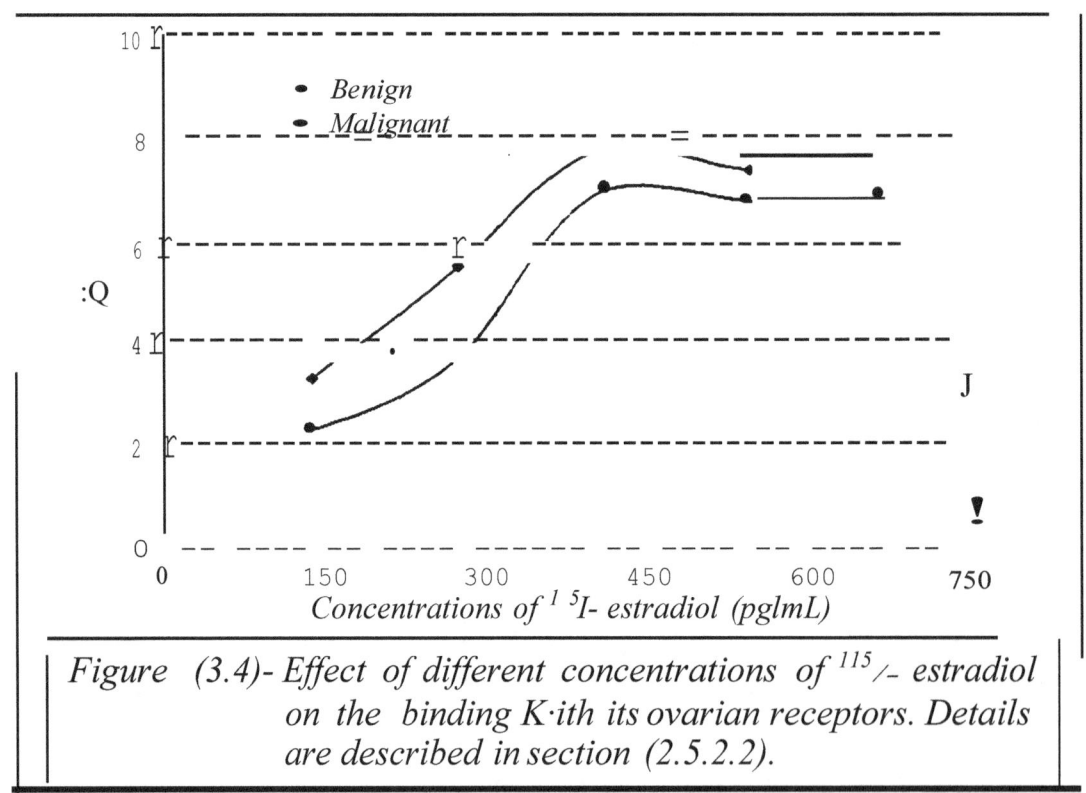

Figure (3.4)- Effect of different concentrations of ^{115}/- estradiol on the binding K·ith its ovarian receptors. Details are described in section (2.5.2.2).

3.3.2.3. The Effect of different pH on ^{125}1-estradio/ binding with its receptors i11 ovarian tumors ho1nogenate.

The effect of pH on the specific binding of ^{115}1-estradiol with its receptors was investigated. Figure (3.5) shows that the optimum pH was found to be 7.6 for the binding of benign and malignant receptors. The same figure shows a decreasing in specific binding percent· at the pH higher or lower than the optimum pH. These results indicate that the binding was pH-dependent and the shift in the pH of the environment may affect the properties of the macromolecules involved in the binding. This effect includes the induction of protonation-deprotonation processes occurring with the ionizable groups of the amino acids present in the binding domain of these macromolecules [11481]. These results are inconsistent \\-'ith the previous work on the uterine homogenates < [1491].

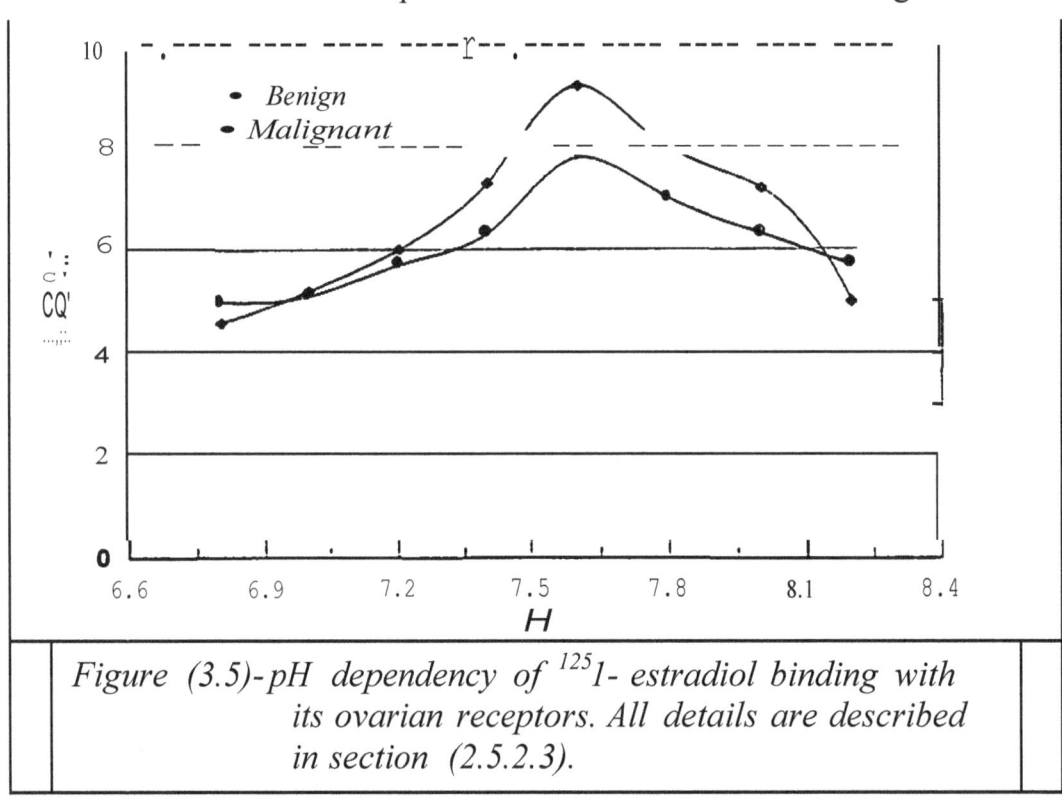

Figure (3.5)-pH dependency of ^{125}1- estradiol binding with its ovarian receptors. All details are described in section (2.5.2.3).

A previous work on the ovarian suggested that the optimal binding occures at pH 7.:> [91], wh"le another work reported that the pH 7.4 is the optimum [162, 1261]. The difference in the pH range may due to the difference of the homogenates source and to the binding condition used.

In view of these results, the pH of the buffer used in all subsequent experiments was adjusted to pH 7.6 for benign and malignant tumors.

3.3.2.4. *The choice of 1nost appropriate incubation time of ^{115}1-estradio/ bindillg to its receptors in ovarian lllllllOr homogenate*

The choice of the most appropriate incubation time was investigated by incubating cytosolic fractions of benign and malignant ovarian tumors homogenate at 4°C for different times, with and without the addition of unlabeled estradiol. fiQure (3.6) shows that the s ecific bindinQ of ^{125}1-estradiol to its receptors was maximal at (18 and 20hrs) for benign and malignant tumors, respectively. These results suggest that the binding was time dependent.

According to the results obtained in this analysis the rime incubation in all subsequent experiments was (18 and 20hrs) for benign and malignant tumors, respectively.

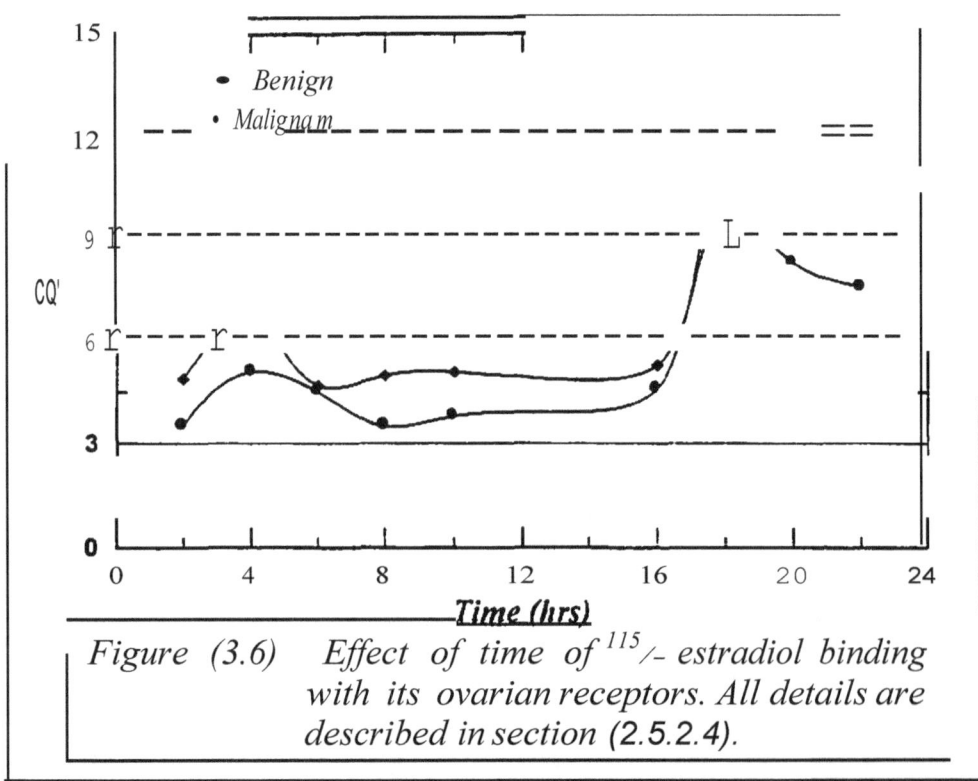

Figure (3.6)   Effect of time of ^{115}/- estradiol binding with its ovarian receptors. All details are described in section (2.5.2.4).

3.3.2.5. Tlte Effect of Temperature on tlte binding o(^{115}/-estradiol to its receptors in ovarian tumors

Temperature dependency of the association of $^{1\ 5}$1-estradiol to its receptors was investigated. C losolic fractions of benign and malignant ovarian rumors were incubated at different temperatures {4-45°C). Figure (3.7) revealed that the specific binding of ^{115}1-estradiol to its receptors was maximal at (37°C and *25°C)* for benign and malignant tumors. respectively. The specific binding \\·as decreased as temperature increase after maximal value of binding. The loss of binding activity may be due to degradation of the receptor molecules o:- leading to the irreversible dissociation of the estradiol-receptor complexes'!:: . ּ\ccording to these. 3*T'C* was used in all the subsequent experiments for **beni** and malignant tumors.

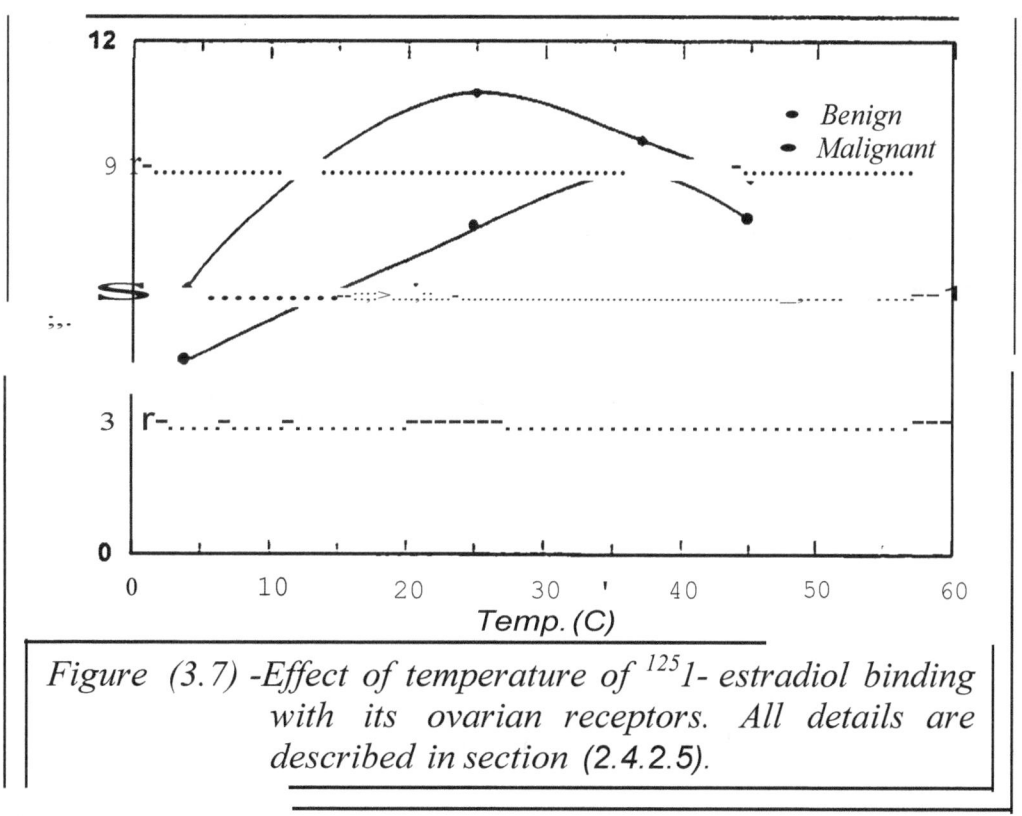

Figure (3.7) -Effect of temperature of ^{125}1- estradiol binding with its ovarian receptors. All details are described in section (2.4.2.5).

3.3.2.6. I}le effect of different halides on the binding of ^{125}1-estradiol to its receptors in ovarian tun1ors

Different halide of sodium in a concentration of (0.1M) were in,·estigated to study their actions on the binding of ^{1} ^{5}1-estradiol with its receptors in benign and malignant ovarian tumor Figure (3.8). The sodium halides in the incubation mixture of benign and malignant receptors induced activation of the percent of specific binding according to the following sequence:

$$aF \cdot > NaCl > NaBr > al$$

The sequence coiTesponds to the decrease of ionic radius, these results are m agreement with those obtaind by Melander and Hor\'ath (1977ı. The same researchers reported t at the effect of halide salt type on hydrophobic interactions is quantified by its molar surface tension increment (MSTI), which is a measure of the increase in surface tension by the salt, also they found that halides with higher (MSTI) values ,.,·ilJ strengthen the hydrophobic L··lteractions while halides with lower (l\.1STI) values reverse this effect.

Thus the dependence of the eA'lent of specific binding of ^{125}1-esuadiol with its receptors in benign and malignant tumors on (MSTI) ,·aiue of the corresponding halide further implicates the low inYolvement ofhydrophobic forces m maintaining the stability of ^{115}1-estradiol-receptor complexes formed[4151].

0!

3.3.2.7. The effect of monovalent and divalent cations on the binding of ^{125}/-estradiol with its receptors in benign and malignant ovarian tumors homogenate.

Figure (3.9) shows the effect of monovalent and divalent cations on ^{1} ^{5}1-estradiol binding with its receptors in benign and malignant ovarian tumors homogenate. Monovalent cations appeared to enhance the binding according to the following order:

$$LiCI > \mathbf{C1} > KCI$$

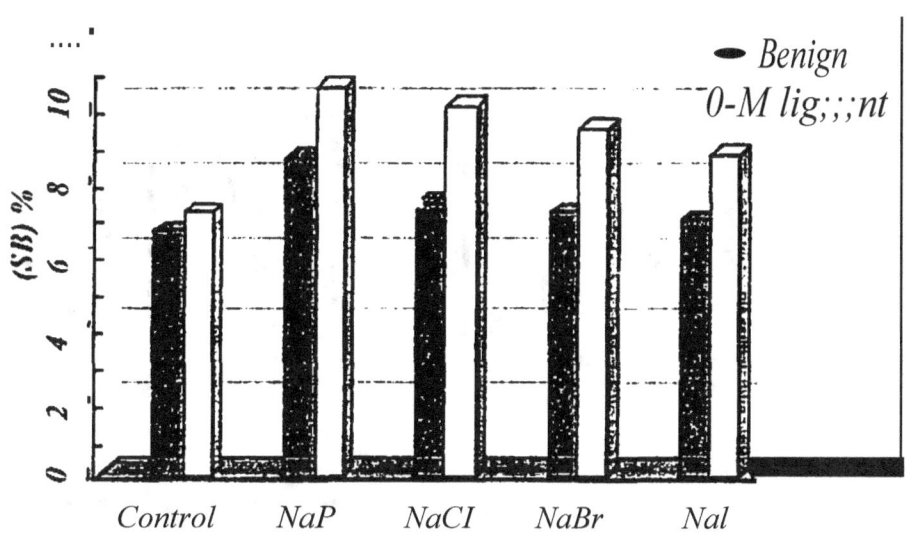

Figure (3.8) -Effect of different halides on the extent of ^{125}/ estradiol binding with its ovarian receptors. All details are described in section (2.5.2.6).

For benign and malignant ovanan tumors, while CsCI decreased the binding for the tow groups of tumors.

DiYalent cations increased the (SB%), while ZnCh decreased the (SB%) for benign and malignant ovarian tumors homogenate. The results indicated that the estradiol binding process is sensitive to the presence of cation.

From the results illustrated in Figure (3.9) it is suggested that these salts may provide some conformational changes in the estradiol receptors and the charged groups of the binding domain of these receptors that hinder maximal binding ns:!. [153].

The inhibiting effect of Zn *(II)* ions on the estradiol binding to its receptors in these results are in agreement with those of other authors who found in their experiments that Zinc was capable of binding with specific sites on steroid receptor molecule and then inhibiting the steroid binding < [154] . [1551] .

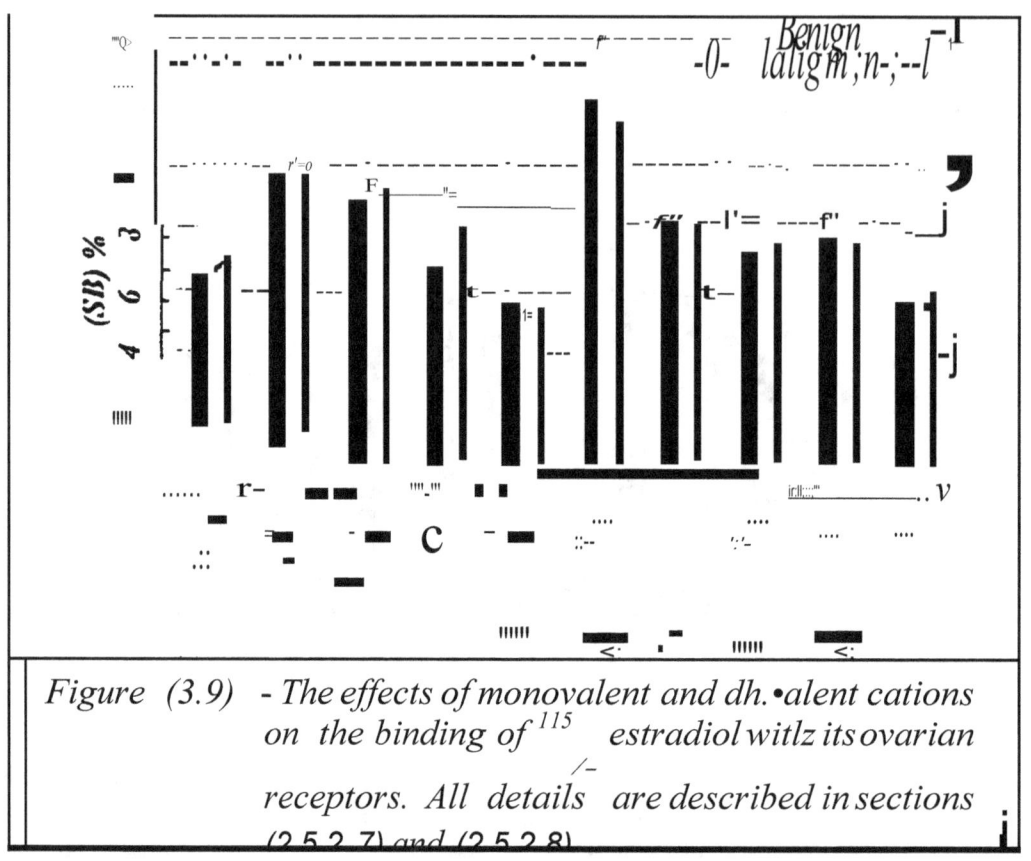

Figure (3.9) - The effects of monovalent and dh.•alent cations on the binding of [115] estradiol witlz its ovarian receptors. All details are described in sections (2 5 2 7) and (2 5 2 8)

3.3.2.8. The effect o(polvethvlene glvco/ fPEG-10000) on the binding o(^{125}/-estradiol with their ovarian tuJnors receptors

Figure (3.10) shows that the incubation of [151]-estradiol with its benign and malignant ovarian tumors receptors in the presence of PEG ranging in concentration from (2-$i0$) resulted in a gradual decreasing of the binding. PEG at 10% concentration was shown to decrease the binding 19% for benign ovarian rumors and 20% for malignant ovarian rumors.

The results indicate that the PEG was able to decrease the (SB%,) of both benign and malignant o·varian tumors, due to its ability to precipitate the protein molecules in the incubation medium so that lead to decrease the interaction between [125]1- estradiol and its receptors.

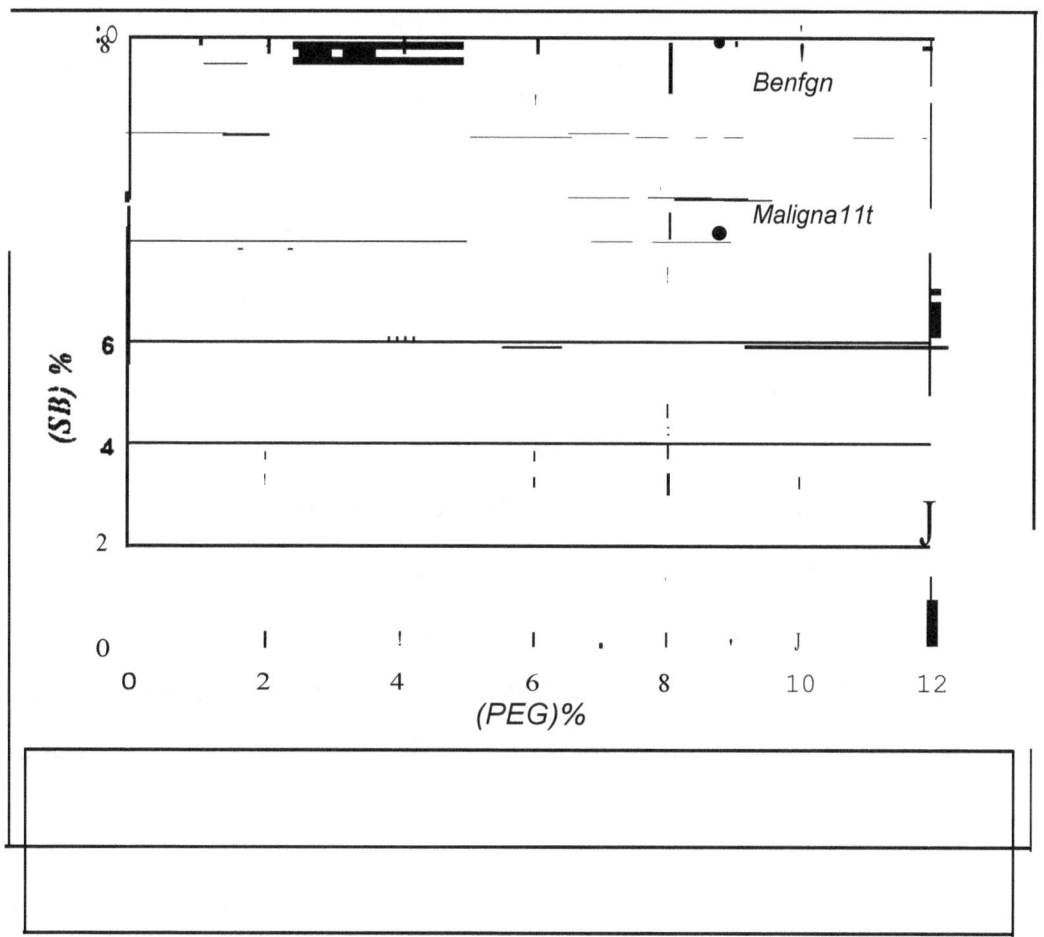

Figure (3.10)- Effect of different concentrations of PEG 10000 (in added TEMG buffer) on the resulted specific binding of ^{115}1- estradiol with its ovarian receptors. All details are described in section

3.3.2.9. Tlze effect of different urea concentration on tile binding of ^{125}/-estradiol to its receptors in ovarian tu1nors.

Different urea concentrations were investigated to study their action on the binding of ^{125}1-estradiol with its receptors in benign and malignant ovarian tumors as shown in Figure (3.11). These experimental results showed that the addition of urea (concentrations ranging from 0.25-4M) resulted in a gradual dissociation of the complex. The urea at 4M was able to dissociate about 37o/o of benign complexes and about 34o/o of malignant complexes, this may be attributed to the effect of urea on the hydrophobic forces participitating in the association of protein molecules.

3.3.2.10. ompetitive effect of different steroid on tlze binding of ^{125}/-estradiol to its receptors in ovarian tumors

The effect of different competitors on the binding of ^{125}1-estradiol to its receptors in benign and malignant ovarian tumors was in, estigated. Figure (3.12) shows that the binding of ^{5}1l-estradiol to its receptors was effecti\"e)y inhibited by unlabeled estradiol, and the competition of progesterone is more than that of testosterone on the binding of ^{125}1-estradiol with its receptors.

The amount of the fixed binding remained. in the presence of competitors at the concentration of maximum displacement was considered to represent the non specific binding [1156]. The displacement of tracer estradiol binding confrrms the specificity of the receptors, which is one of the fundamental criteria of the true receptors [1157, 158].

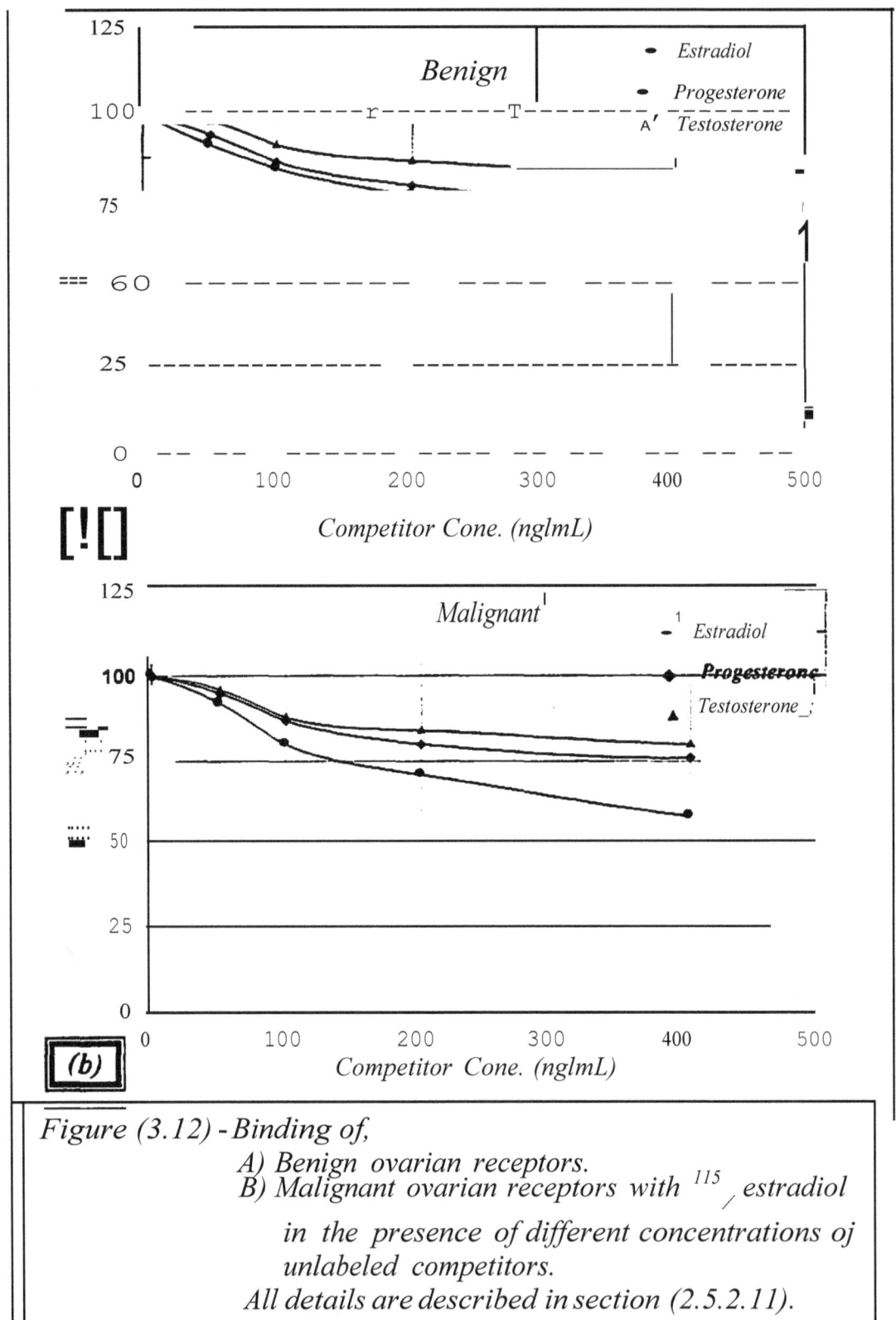

Figure (3.12) - Binding of,
   A) Benign ovarian receptors.
   B) Malignant ovarian receptors with ^{115} estradiol
      in the presence of different concentrations oj
      unlabeled competitors.
   All details are described in section (2.5.2.11).

3.4. The kinetic and thermodynamic studies o(^{125}/-estradiol binding with its receptors in benign and 111alignant serous ovarian tumors /zo1nogenate.

3.4.1. The kinetic studies o(^{125}1-estradiolbinding with its receptors in benign and malignant serous ovarian tunzors homogenate.

Figure (3.13 a and b) shows the time course ofthe formation ofC^{15}I-estradiol-receptor) complex for benign and malignant o\·arian tumors, respectiYely. at different temperatures (4. 2.5. 37 and 45°(). The results oftime course panems at different temperatures revealed t at the binding of ::..:;!-estradiol to its receptors in O\'arian tumors is a temperature and time dependent process with a maximum binding occurs at 37°C with 18 hr for benign receptors and at 25°(with 20 hr for malignant receptors.

3.4.1.1. Determination of estradiol receptors concentrations and the a(finitv constants of ^{125}/-estradiol association with its receptors in benign and malignant ovarian tumors

C 1osolic estradiol receptor concentrations and the affinity constants have been measured in benign and malignant O\'arian tumors. The experiment was carried out at the optimal conditions. which were obtained in preYious experiments and was repeated at different temperatures (4, 25, 37 and 45cC). Scatchard plot analysis gave a straight line as shown in Figure (3.14 a and b) at each temperature indicating the presence of only one species of receptor site. or more but with the same affinity and number ofbinding sites.

The results are summarized in Table *(3.2)*. In the optimal conditions the estradiol receptors concentrations were (116fmollmg protein and 145fmol·mg protein) for benign and malignant ovarian rumors, respectively, These results are inconsistent with those reported previously, that the estradiol receptor concentration was hi __her in the malignant lesions than that ofbeniQll <BO>.

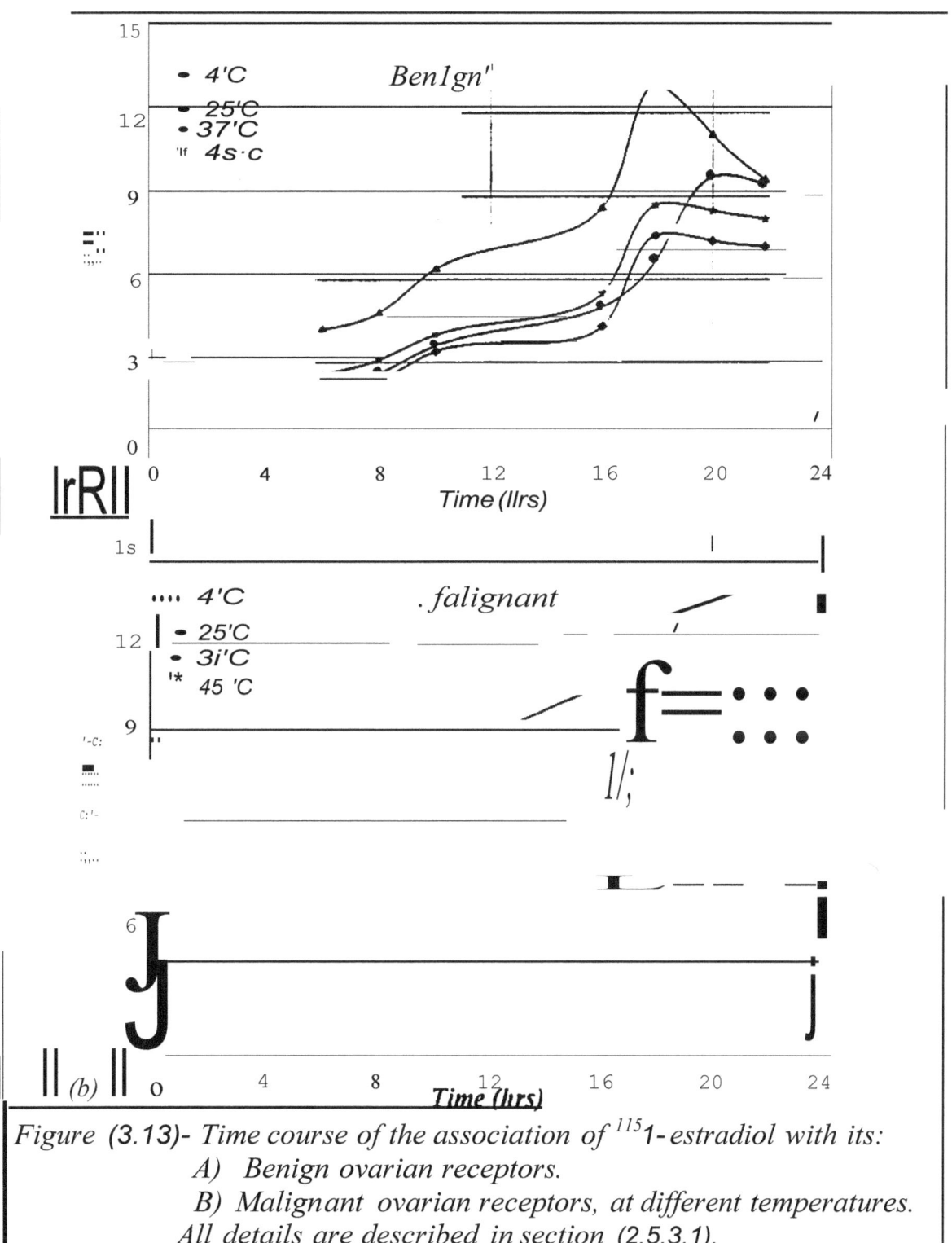

Figure (3.13)- Time course of the association of ^{115}1- estradiol with its:
   A)   Benign ovarian receptors.
   B)   Malignant ovarian receptors, at different temperatures.
   All details are described in section (2.5.3.1).

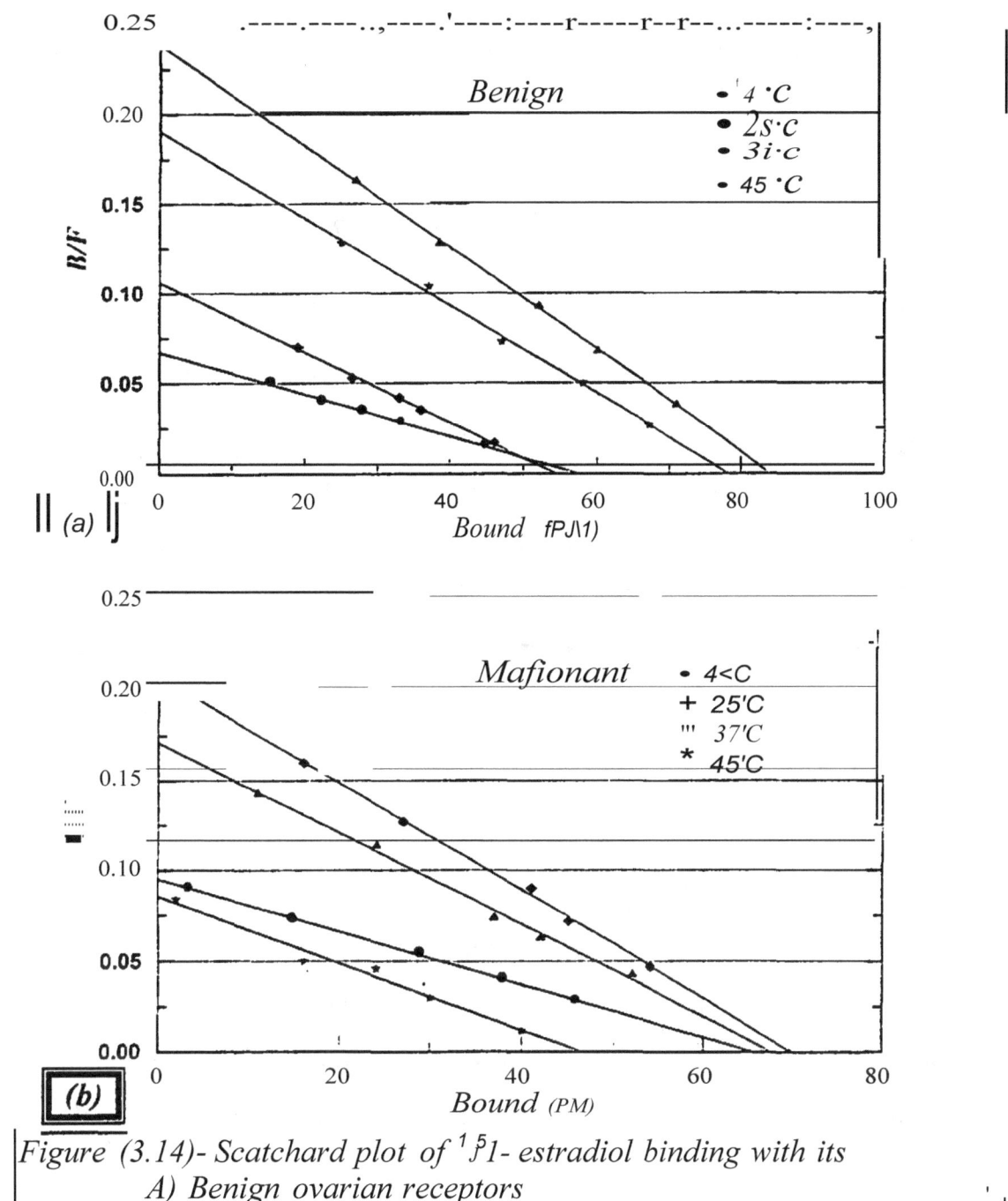

Figure (3.14)- Scatchard plot of $^{15}1$- estradiol binding with its
  A) Benign ovarian receptors
  B) Malignant ovarian receptors atfive different temperatures.
  All details are described in section (2.5.3.2).

Mazia Lanna (1984) reported that the presence of steroid recepto:-s in malignant ovarian tumors can possibly be used as an indicator to hormone dependency of these rnaltgnancJes l)JfJ,. Prev10us stu^{d}tes on t^{h}e b$^"$m d$^"_{m2}$ o^f lablled estradiol with. its receptors in o\·arian tumors suggested that t.:te maximum binding sites (Bmax) of estradiol receptor varied between 11 and 397 fmole/mg protein and the $Ki < 1 O^{.9}$ M [059] [160].

Table (3.2) - The kinetic parameters of ^{125}1- estradiol binding to its receptors in benign and malignant ovarian tumors. All details are described in section (2.5.3.2).

Temp. (CC)	BeniKn Binding capacity ({mole! lmg protein)	$Ka \times lo'$ ("lf^{-1})	$Kd \times lo^{-''}$ (l•f)	Malignant Binding fcapacity (fmole'j lmg protein)	$lxax JO'$ (M^{-1})	$Kd \times Jo$-Jo (.\1)
4	79.16	11.6863	8.557	135.4	114.5067,	6.893
25	75	19.4857	5.131	145.8	29.659	3.37
37	116.6	28.28	3.536	141.6	25.219	3.965
45	108.3	'24.3447	4.107	93.7	19.3904:	5.157

3.4.1.2. Determination of kinetic parameters o(^{125}/-estradiol binding with its crtosolic receptors in benign and nzalignant ovarian tumors

The time course of $^{1 5}$1-estradiol binding to its receptors in benign and malignant ovarian tumors was carried out to describe the kinetic parameters of the binding. The simplest proposed model representing the interaction of $^{1 5}$1-estradiol with its receptors could be expressed by the following equation:

$$^{125}\text{I-estradiol} + R \underset{k_.)}{\overset{k\text{-}J}{\rightleftharpoons}} {}^{12:'}\text{I-estradiol-R}$$

Where

$k+_I$ is the rate of the association of ^{125}1-estradiol with its receptors

$k.J$ represents the rate of the reverse reaction of the dissociation of the complex formed under the same conditions

At equilibrium;

$$k_a = \frac{1^{115}I\text{-}estradiol\text{-}Rj}{1^{115}I\text{-}estradiol\,II\,Rj} \quad \dots\dots\dots\dots\dots\dots\dots\dots\dots \quad (1)$$

$$k_d = \frac{1^{115}I\text{-}estradiol\,II\,Rj}{f^{15}I\text{-}estradiol\text{-}Rj} \quad \dots\dots\dots\dots\dots\dots\dots\dots\dots \quad (2)$$

Thus;

$$k_D = \frac{1}{kd} - \frac{k}{k\_/} \quad \dots\dots\dots\dots\dots\dots\dots\dots\dots\dots\dots\dots\dots \quad (3)$$

Where

Ka is the equilibrium constant of the association (the affinity constant)

Kd is the equilibrium constant of the dissociation of $^{1\,5}$1-estradiol-R complex.

The Yalues of ka and maximal binding capacity (Bmax) were calculated from Scatchard plot at four different temperatures as in Figure (3.14) and Table (...)_').

Results m Table *(3.2)* show that ka value at 37°C is higher than that at (4, 25 and 45°C) for benign tumors, hile ka value at 25°C is higher than that at (4, 37 and 45°C) for malignant rumors.

The values of calculated by using equation (3) show that the lowest \alue of ^{115}1-estradio]-receptor complex occurs at 37°C for benign receptors and at 25°C for malignant receptors. Also, it was found that the affinity of estradiol to its malignant receptors was greater than that of benign receptors in temperarure from 4 to 25°C, while in temperatures from 37 to 45°C, the affinity of estradiol to its benign receptors was greater than that of malignant receptors.

For verification of the order of the reaction by using the data from the time-course of ^{115}1-estradiol binding with its receptors from benign and malignant ovarian·tumors homogenate at four temperalure, and Since in the

cases of this work, the percent of the specific binding was small, and most of the [1]1-estradiol remained free and only a small fraction of the total concentration of [1][5]1-estradiol is bound even at equilibrium (pseudo-first order conditions), so that the following equation $_{<1}$[71] could be used in order to fit the data of the first-order kinetics:

$$_{ln}(HR \; lHR), = t.k... \quad\quad \dots\dots\dots\dots\dots\dots\dots\dots\dots\dots\dots\dots \quad (4)$$

Where

fHR)e is the concentration of e^2;I-estradiol-receptor) complex formed at equilibrium,

(HR), is the concentration of ([1][5]1-estradiol-receptor) complex formed at time (t) and

Kobs is observed value of first rate constant.

Figure (3.15 a and b) shows that the plotting of $ln (\dfrac{HRl,}{HR \; - \; HR})$ against time *(t)* which gives a straight line with a slope equal to the O<obsJ in hr-[1].

The association rate constant k-$_1$ "/as calculated from the follo,ving formula [1141]:

$$kobs. =_{k+l} \dfrac{(H)r(R)r}{(HR)} \quad \dots\dots\dots\dots\dots\dots\dots\dots\dots\dots\dots\dots\dots\dots\dots\dots\dots\dots \quad (.))$$

Where

(H)r is the total concentration of ^{125}1-estradiol and

(R)r is the total concentration of receptors of benign or malignant o\·anan tumors homogenate.

The halflife time of association (t$_{112}$)ass., which represents the time needed for the formation of half amount of the complex at equilibrium, was determined

from the concentration of the complex at equilibrium and the time course cun·e.
While the half-life time of dissociation ($t_{1/2}$)diss. was determined from:

$$(r t.)._{dw} = \ln \frac{!}{k-1} = \frac{0.693}{k-1} \quad\dotfill \quad (6)$$

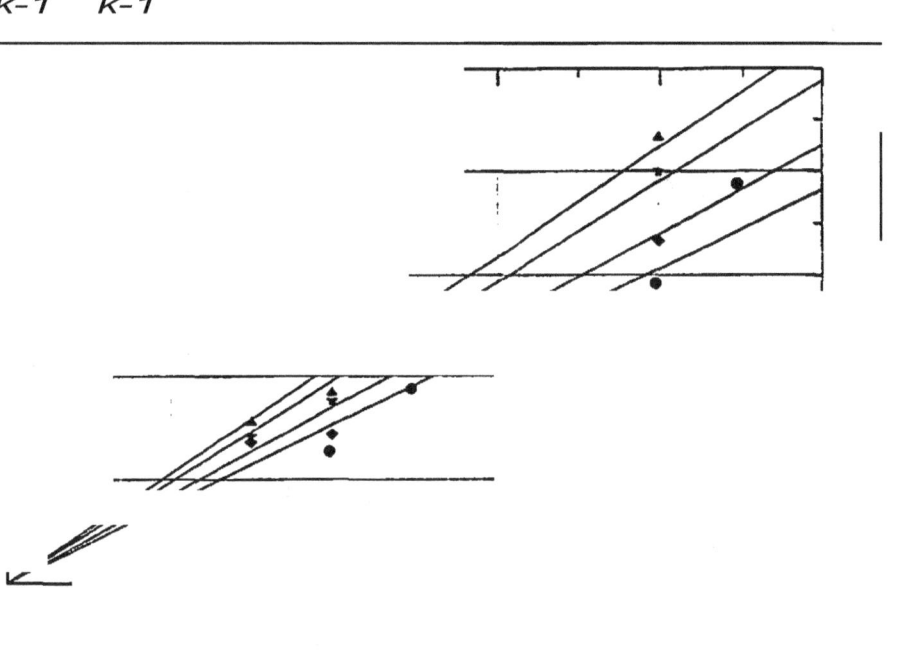

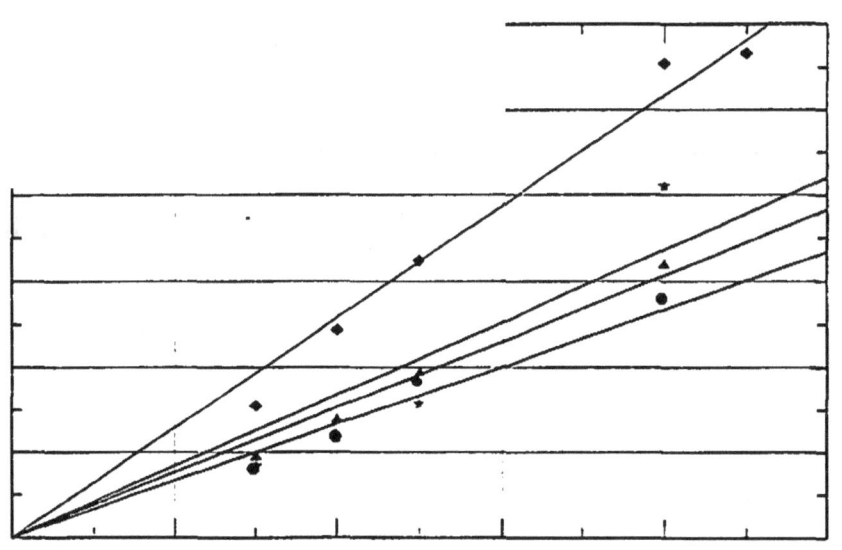

(b)

The results revealed that the association rate constant k_{-1} at 37°C was higher than that of at 4OC by approximately 1.3 folds for benign tumors while:- the k_{-1} value at 25°C was higher than that of 4°C by approximately 3.5 folds for malignant tumors

Table (3.3) shows the k-1. k..t. (t1 :)a... and (t1 :)diss for benign and malignant tumors.

Table (3.3) - The effect of temperature on the kinetic parameters of estradiol
bindin g to Its receptors in ben1gn and nzalignant ovanan tumors

Benign tumor							
Temp. (°C)	Binding capacity (fmole/mg protein)	k_a $(M^{-1}) \times 10^8$	k_d $(M) \times 10^{-1c}$	K_{-1} $(hr^{-1}.M^{-1}.10^8$	K_{-1} (hr^{-1})	$(t_{1:})_{ass}$ (hr)	$(t_{1/2})_{dis}$ (hr)
4	79.16	11.6863	8.557	1.127	0.0964	16.6	7.18
2*	75	19.4857	5.131	1.149	0.0589	11.7	11.76
37	116.6	28.28	3.536	1.507	0.0532	10.1	13
45	108.3	24.3447	4.107	0.801	0.0329	11	21
Ma/ig17,ant tumor							
Temp. (OC)	Bilzding capacity nmof.'mg protein)	li_{11} $('.'d^{-1}) \times 10^8.'M$	lid $'>\times 10^{-lt}$	K_{-1} $(hr^{-1}.M^{-1}.10^8$	K (llr_{-1})	$(t':Ls$ $(hr^>$	$(t_m tu$ $'llr^>$
4	135.4	14.5067	6.893	0.691	0.0476	15.2	14.5
25	145.8	29.659	3.37	2.432	0.0819	9.1	8.46
37	141.6	25.219	3.965	1.197	0.047	11.6	14.7
45	93.7	19.3904	5.157	1.367	0.07	12	9.9

3.4.2. Tlze thernzodvnamics of the binding of ^{115}I- estradiol to its receptors in benign and malignant ovarian tumors

3.4.2.1. Thermodvnamic parameters o(standard state

Figure (3.16) represents the dependence of the equilibrium binding constant (affinity constant) for the binding of ^{125}1-estradiol to its receptors in ovarian tumor homogenate on the temperature (Van't Hoff plot).

The results indicated that $L:lH^0$ in general had small values and nearly close to zero, their positi\·e sign ascertain that the reaction was nearly endothermic. The $L:l!f>$ value in the case of benign receptors was higher than that in the case of malignant receptors, so more energy is needed in case of benign receptors for the binding to occur.

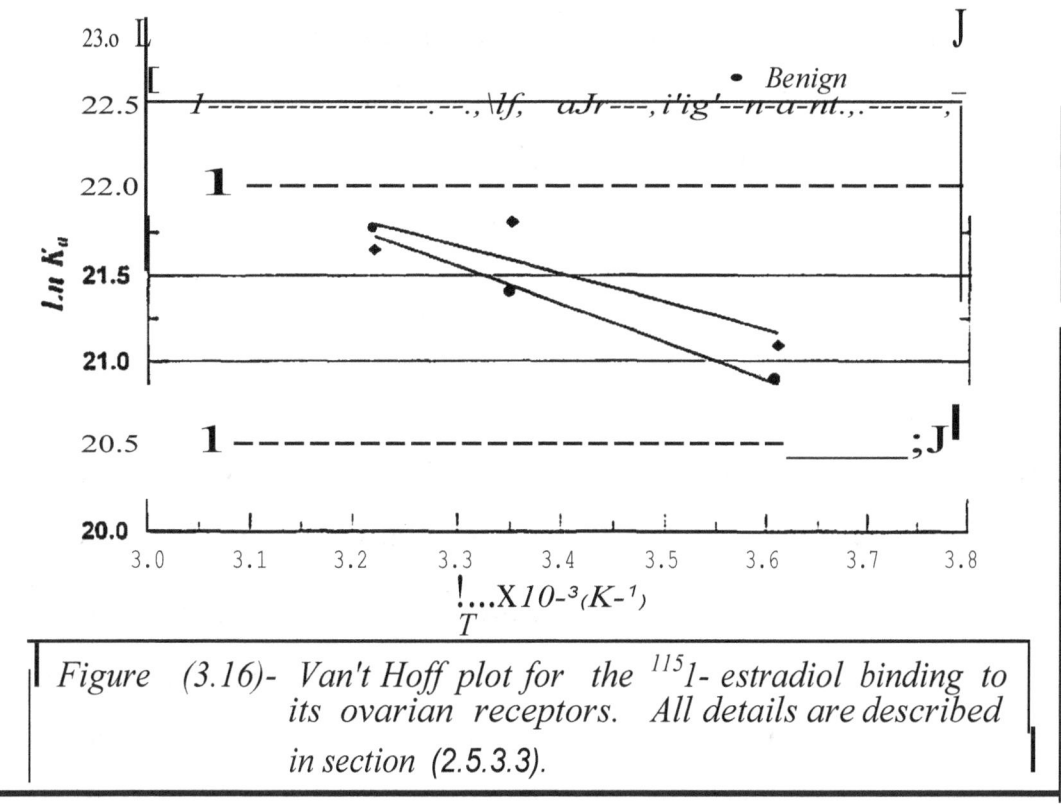

Figure (3.16)- Van't Hoff plot for the ^{115}1- estradiol binding to its ovarian receptors. All details are described in section (2.5.3.3).

The negatt\·e values of Go retlect the stability of the complex hence. the high affinity of the reactants. The high negative values of Go for the binding rea :ions are controlled by high positive D.S" values as shown in Table (3A). So. our system is characterized b"·the sole contribution of So to the stability of the complexes forme<L while has little or no effect.

A high value of positiv . $L:lS^0$ suggests that the reaction spontaneity \';as enrropically driven. Entropy was the driven force for the occurrence ofthe binding reaction. This indicates that the hydrophobic interactions played an important role in stabilizing the complex $_1$[162]$_1$.

Table (3.4) showed that LlH^0 and $.65°$ values for the binding of $^{115}1$- estradiol to its benign ovarian receptors were higher than those for malignant receptors, this means that binding of $^{115}1$- estradiol with its benign receptors needs more energy, and the complex formed possessed a less ordered structure than the reactant species.

The small positive $.6rr'$ may indicate a favorable interaction berv.:een groups within both estradiol and its receptors. These include the non-covalent interactions. which are fundamentally electrostatic in nature such as charge-charge interactions, which occurs in both estradiol and its receptors in ovarian rumor homogenate, other types of interactions include charge-dipole. dipole-dipole, charge-induced dipole. dipole-induced dipole and hydrogen bond. The sum of these types of interactions can yield some stabilization to the folded structure of the complex. So. the negative value of $.D.G^0$ showed that the o\·erall reaction was energetically favorable in the direction of complex formation.

The results listed in Table (3.-+) indicate that the reaction spontaneity for benign receptors was smaller than that for malignant receptors at optimum temperature ^{163}. IM). Also, it was concluded that short range interactions including

Van der Waals· interactions. protonations and hydrogen bond formation were more :rrtportant factors in stabilizing the benign receptor complex than that of malignant receptors.

Table (3.4) - Thermodynamic parameters at standard state of estradiol binding to its receptors in benign and malignant ovarian tumors. All details are described in section (2.5.3.3).

3.4.2.2. Tltermodynamic parameters of Transition State

The transition state theory proposes that the interaction of [125]1-estradiol and its receptor lead to the formation of an activated complex (transition state), then the formation ofthe final product:

$$m \text{ } l\text{-}estradiol + R \text{ } t\text{.}^{\bullet} - estradiol\text{-} R\}\text{-} + \text{ } ^{1}\text{.}sl\text{-}estradiol\text{-}R$$
$$Allacril\text{:}ated comple.\text{:-}c \qquad Final \text{ } product$$
$$(Transition \quad state)$$

The transition state thermodynamic parameters (.aH*, G·, S· and Ea) could be determined from Arrhenius equationand the kinetic constants.

Figure (3.17) shows the Arrhenius plot of the In k-1 against 1/T values (the dependence of the association rate for the binding of [125]1-estradiol with its receptors in benign and malignant ovarian tumor homogenate on temperature).

Table (3.5) shO\VS the values of thermodynamic parameters of transition state (Mf*, G·, .6.S* and $E_{3)}$. The value ofEa determined from Arrhenius plot represents the apparem energy of activation of the binding reaction and the requireed energy to overcome the energy barrier of the transition state for the formation of [^{125}1- Estradiol -R] complex.

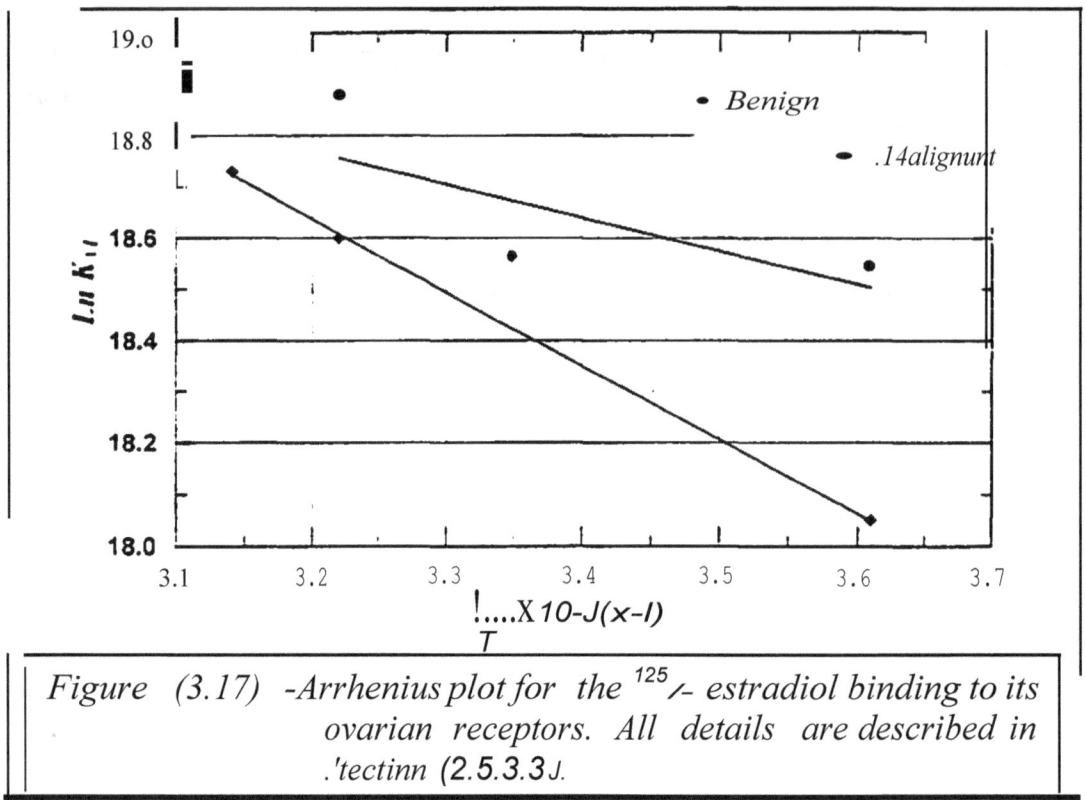

Figure (3.17) -Arrhenius plot for the ^{125}/- estradiol binding to its ovarian receptors. All details are described in .'tectinn (2.5.3.3J.

The high positive value of .o.G" indicated that the formation of an activated C^{25}I-Estradiol-R] complex was a non spontaneous process and required a lot of energy (equal to Ea) to overcome the transition state energy barrier and giving the final product. whereas the high negative D.S• revealed that the activated complex had a more ordered structure than the reactant species (D.S" < 0). Tne ₍ⱼⱼ₎ positive values of D.G. is mainly anributed to the decrease in entropy of the transition state <..lS.<O). In addition. the positive value of D.H• shows that the heat content of the activated complex is more than that of isolated species [1165].

The results in Tabl · (3.5) shows that the values of Ea and D.H. for the binding reactions of estradiol with its malignant ovarian receptors were more than that in the case of benign receptors and the Ea value for malignant receptors was greater than that of benign receptors by a factor of 2.2. Therefore, the binding reaction of estradiol with benign receptors was easy to occur when compared with the same reaction with malignant receptors. The negative values of D.s• for malignant receptors ......,.ere slightly greater than that of benign receptors, so it was concluded that hydrophobic interactions may play an important role in stabilizing the malignant receptor activated complex formed.

Table (3.5) - Thermodynamic parameters at transition state of ^{125}I- estradiol binding to its receptors in benign and malignant ovarian tumors. All details are described in section (2.5.3.3).

Temp. (CC)	Benif!n receptors				Ma/if!nant receptors			
	Ea (kJ!/mole)	MI* (kJ/inole)	!IG* (kJ/molel	I:JS* (Jimole.K)	Ea lkJ/mole)	M* (kJimolel	IIG* if.kJ/mole)	IIS* (Jimole.K)
4	5.39	+3.08	+24.975	-79.04	11.93	+9.62	+26.104	-59.5
25	5.39	+2.91	+27	-80.83	11.93	+9.45	+25.141	-52.65
37	5.39	+2.81	.J-27.493	-79.62	11.93	+9.35	+28.086	-60.43
45	5.39	+2.74	+29.935	-85.51	11.93	+9.28	+28.534	-60.54

Determination of the thermodynamic parameters of the binding reaction using equilibrium data gives an overall idea about the nature of forces controlling complex formation. Comparison of the values of transition state with those of standard state in Table_s (3.4) and (3.5) led us to choose a thermodynamic model shown in Figure (3.18). Our model proposes that the formation of the ^{125}1-estradiol-receptor complex undergo three thermodynamic states. Thermodynamic state A represents the initial energy level of the isolated ^{125}1- estradiol and its receptor. In thermodynamic state B, the two components have come together and mutually penetrated their hydration sphere to form a partially immobilized hydrophobically associated species. Thermodynamic state C represents the fully interacting complex ('.^{5}1-estradiol-R). In step 1 ofthe reaction, the binding of ^{125}1-estradiol to its receptor was associated with positive ilG• value. This indicates that the initial step of the reaction requires input of energy for the system. The negatiYe entropy change ilS• for this step of the reaction reflects the change of the ^{1}.^{5}1-estradiol-R transition complex to a more ordered structure. The positive LUI. value shows that the heat content of the activated complex is more than that of the isolated species. Partial immobilization of the hydrophobicaJly associated complex formed, in Step 1. occurs when isoiated h>'drated species ^{1} ^{5}1-estradiol and receptor interact partiaily so that there is a mumal penetration of their hydration layers to tonn the activated compiex. Thi:; h dwphooic association is •csuit ofth t.ende:1cy of \Water to form a more ordered structure in th!! vicinity of non-polar hydrocarbon groups (cg.• the side: chains of the ::unino adcis phenylalanine. leucine and t.ryptopban) this means that hydrophobic amino acid side chains which \\:ere prev ously accessible to solvent in the isolated species become bu:ied upon compicx tbrmP-tion. fn step *1*.the activated ..:omp!cx panicipates in funher interactBons. giving the fuily interacting compiex (IZ-'Isestradiol-R). It is prop-osed that tbe rbrmation of a (t:: l-cstradiol-R) complex occurs En ttNo steps, the tlrs:: step tabilizcc the complex by hydrophobic interactions and the second step stabilized i{ by sho11 range merncdons such as elcctrosmtie interactions. hydrogen bondLn:; ::md 'ion del' \V.uuls' imcmcrions •1'>f•.

H;:drophobi..: inter..tctions contribute: to the s:abilily of the complex via large decrease in the t!Xcit d entrop chenge (S•<0_), while electroStatic inte:raction.s. hydrogen bonding and V n der \V::a!s' interactions stabilize the complex via high increase in the standard entropy change (ΔS'>0) [168, 169].

The Ulro! c...C;. r.:.lr:'lic dau from this :>tudy indic u lhat th binding :; (. estradiol to its r c'>!ptors are c!ntropically driven and come in ag.reem nt .....ith the concept tbat h ·drophobic interactions ha,_.e an important ro e in the binding. of estradiol to its receptors.

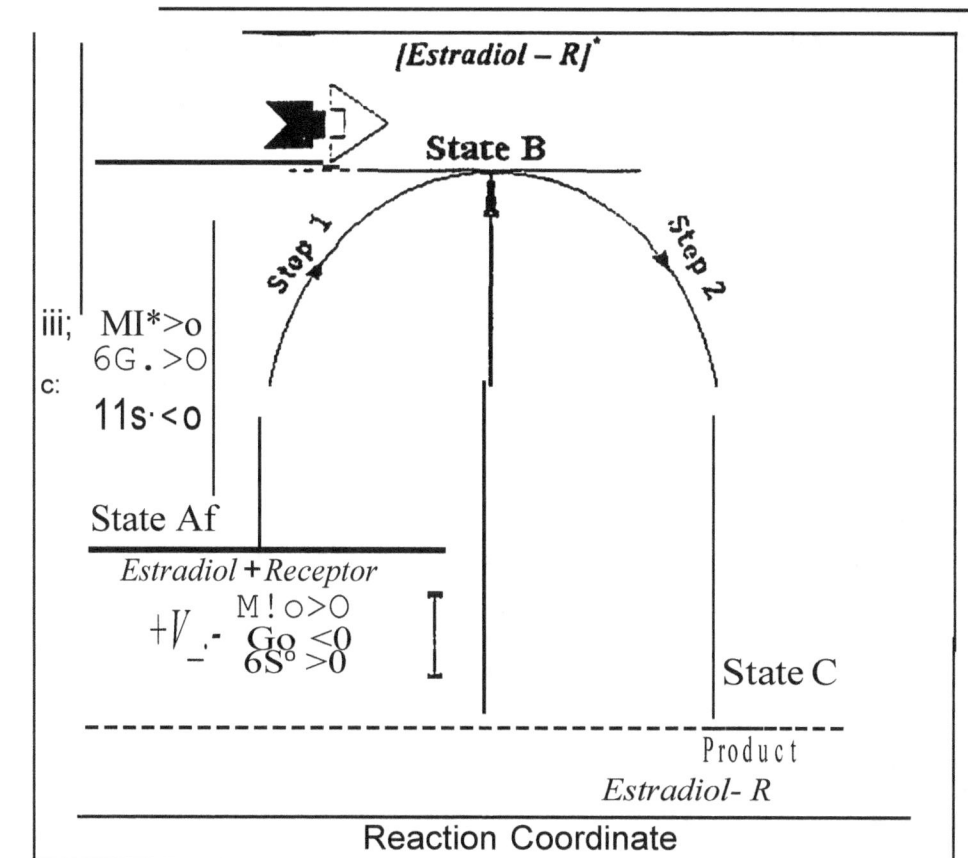

Figure (3.18) - General energy diagram and thermodynamic nwde/ applied to the interaction of binding oj [125]!-estradiol with its receptors in ovarian tu1nors.

3.5. Isolation of cvtosolic estradiol receptors using gel filtration technique₁

Figure (3.19) shows the results of gel filtration technique used to isolate of estradiol receptors from Benign and malignant ovarian tumor homogenate. Benign and malignant homogenates were applied to sephadex G-150 (0.7x23 em) column. The void volume of this column was 7m.L as sho\\ITI in Figure (3.19). The resultant fractions of each homogenate type \vere collected, detected for the binding with [125]1-estradiol as described in section (2.6), pooled, concentrated and then subjected to protein determination as mentioned in section (2.2). This experimcnt revealed the presence of two different eluted components (I and II).

these two components eluted with different elution volume corresponding to their different molecular weights, and with tow absorbance peaks for protein content.

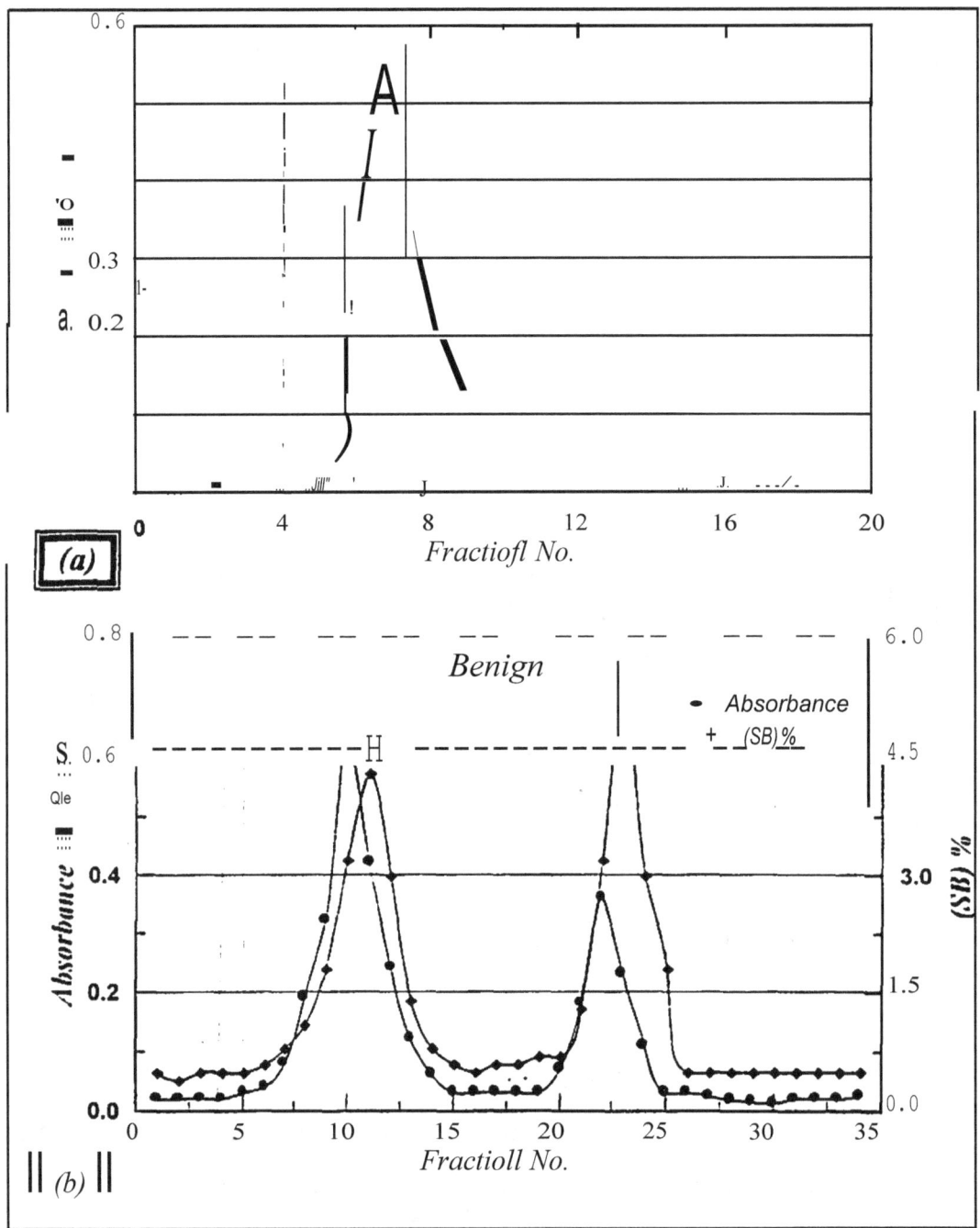

(a)

(b)

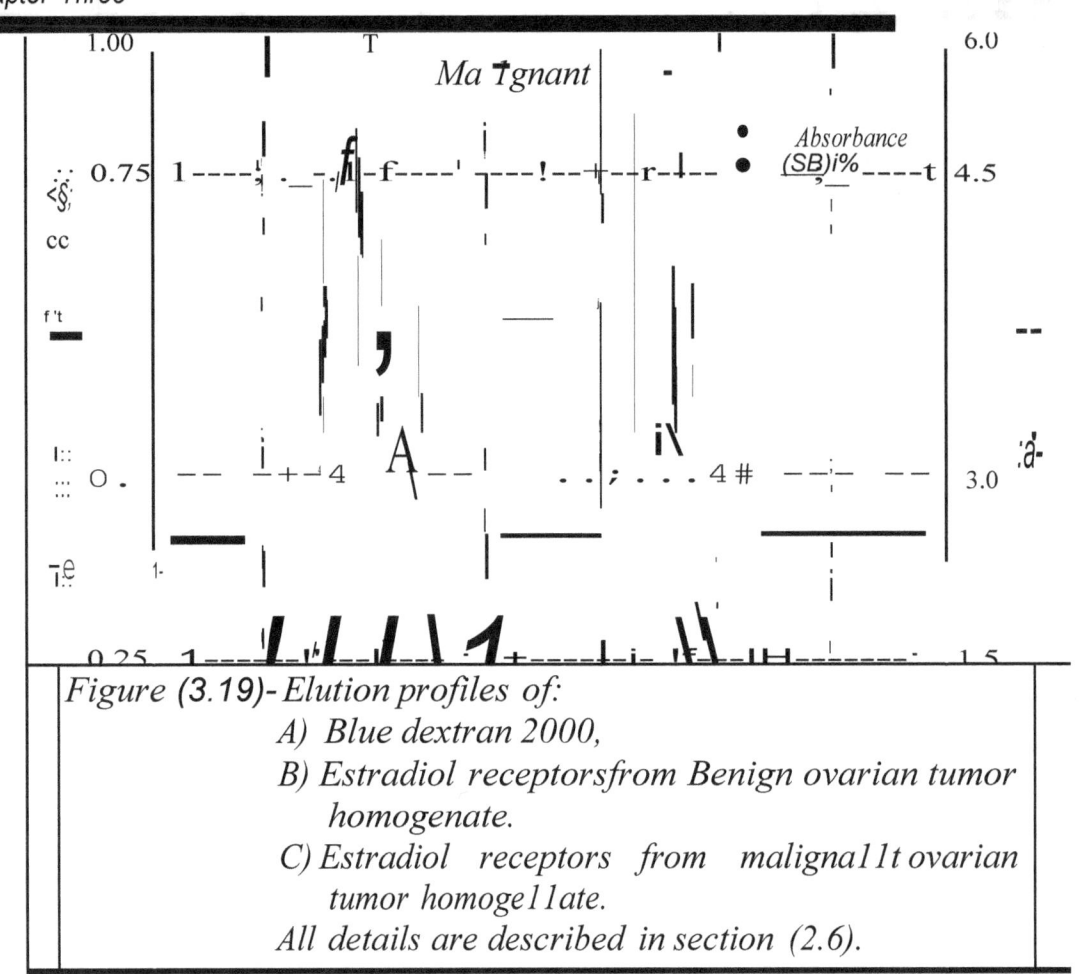

Figure (3.19)- Elution profiles of:
A) Blue dextran 2000,
B) Estradiol receptorsfrom Benign ovarian tumor homogenate.
C) Estradiol receptors from malignal1t ovarian tumor homoge11ate.
All details are described in section (2.6).

From benign and malignant tumors homogenate, the first one (BI& Ml) eluted with about $1.6V_0$ (fraction number (11)), while the second one (BII& l\111) eluted with about $3.3V_0$. (fraction number *(23*)). From these results, it was concluded that t ese components are capable of binding to the estradiol with different affmities and in general receptors type II have higher affinities for the binding than tho e of receptors type (1). Two sbsorbance peaks for protein content were obtained with maximum absorbance in fraction number (10 and 22) for benign receptores and in fraction number (9 and 23) for malignant receptors. Table (3.6) illustrates the isolation parameters for the different estradiol receptor forms isolated by gel exclusion chromatography technique.

3.6. Molecular weight determination

The molecular weight of estradiol receptors was determined by using gel filtration chromatography. The void volume (V) of the column was determined by using blue dextran and found equal to (7mL). Different standard proteins (ferritin(+fOKD), catalase (232KD), aldolase (158KD), and BSA (67KD))were applied separately t column and their elution volume (Ve)were measured Figure (3.20). The (Kav) values for these proteins w·ere calculated, and a calibration curve was plotted between Kav value of the standard against their logarithmic molecular weight Figure (3.21). The molecular weights of nvo estradiol receptor proteins obtained were determined. It was found to be 288KD for component I. and 75KD for component II. These results are in a good agreement \with those reported previously [11561] =

Table (3.6)- Data of estradiol receptors isolated by gel filtration technique. All details are described in section (2.6).

Receptor type	Total proteins (pgm)	Specific aUy boundi2Sl- pu J{pM)	Specific binding Fnwle ^{125}I- estradiol /mt! protein	isolationfold
Crude benign awuian tumors homogenate	360	10.9	15.18	1
BI isolated fraction	175	11.8	62.3	4.1
BII isolaJed fraction		11.8	95	6.25
!Crude Mlllignant qvarian tumors homogenate	110			
isolated	240	15.9	33.1 .99.4	1
iWI fraction	160	31.8		5.:J
		33.8	169	

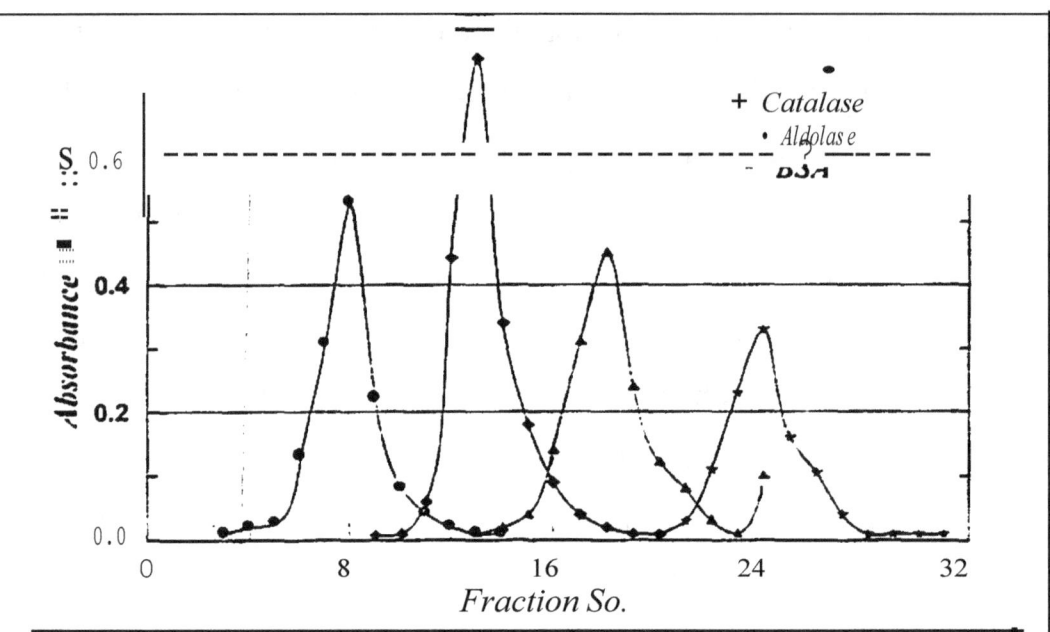

Figure   f3.20)   - Slteplzade..\: G-150 clzrOJnatograplzy of standard proteins,   all   the   details   are   described   in section (2.7).

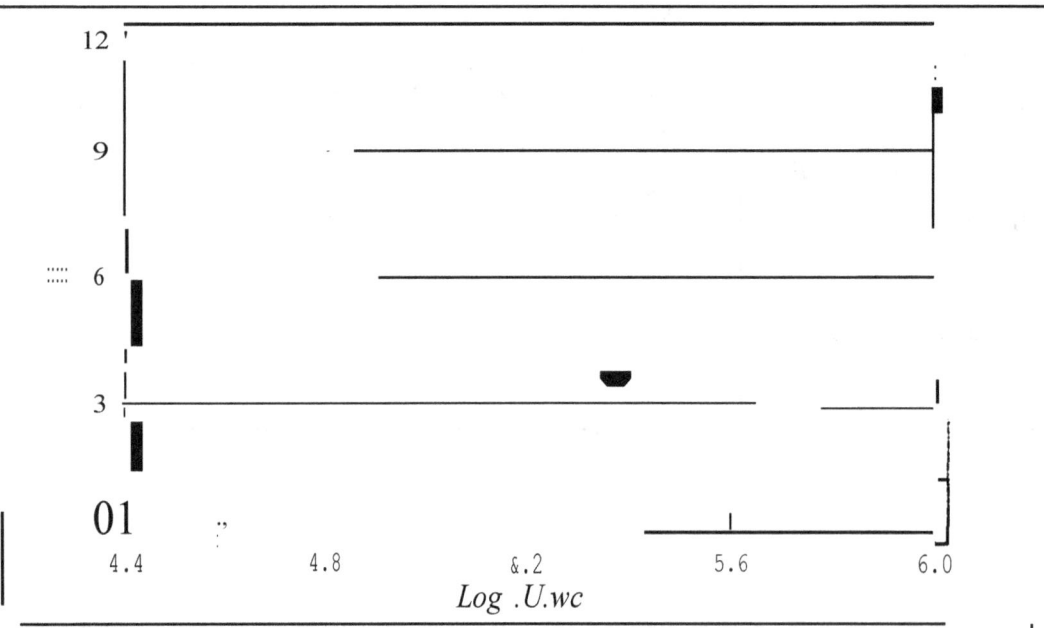

Figure (3.21) Caiibratio11 cun·e for deternzination of Jt/.wt by gel filtration chromatography. All the details are described in section (2.7).

3.7. Spectroscopic studies of different isolated forms of estradiol receptors

3.7.1 The U.V. spectra of isolated estradiol receptors in benign and malignant ovarian tu1nors.

Figure (3.22 a and b) illustrates the U.V. spectra of different isolated forms of estradiol receptors at pH *7.-1-*. The U.V. spectra show that the /'max for the isolated receptor BI is consists of one peak at 206.1 nm, BII- isolated receptor gh·es two peaks at 204.3nm and 273.4nm, \-fl- isolated receptor gives one peak at 209.6 nm and Mil- isolated receptor gives three peaks at 202.8 nm. *223*.2nm and 260.3nm. .-\s a result each estradiol receptor has a characteristic spectrum and can be identified by their peaks 206.1. *204.3 209.6,* 202.8, and

260.3 nm are assigned to phenylalanine residues. while 273.4nm and *223.2nm* are assigned to tyrosine residues ([168.1] ..... .-\lso it \vas found from the Figure *(3.22 a and b)* that tryptophan residues does not occur on the surface of benign and rnal'tgnant receptors [169].

It seems that in BI. BII. MI and :YIII- isolated receptors, all phenylalanine residues seem to· be on the surface of the receptor molecules, exposed to absorbance. On the other hand. Tyrosines of the HI-receptor molecule seem to be on the surface, in BII- isolated receptor. these are located in a way that part of it is on the surface of the receptor molecules and the other parts are buried. \vhile in BI and l'vll-receptors these residues may be completely buried.

3.7.2. Factors affecting the absorption properties of isolated estradiol receptors in beni!!Jl and 1nalignant ovarian tzunors

The absorption spectrum of a chromophore is primarily determined by the chemical structure of the molecule. However, a large number of environmental factors produce detectable changes ir:t Amax and E. Environmental factors such as pH and polarity of the solvent provide the basis for the use of absorption specuoscopy in charac,t.erizing macromolecules[1].

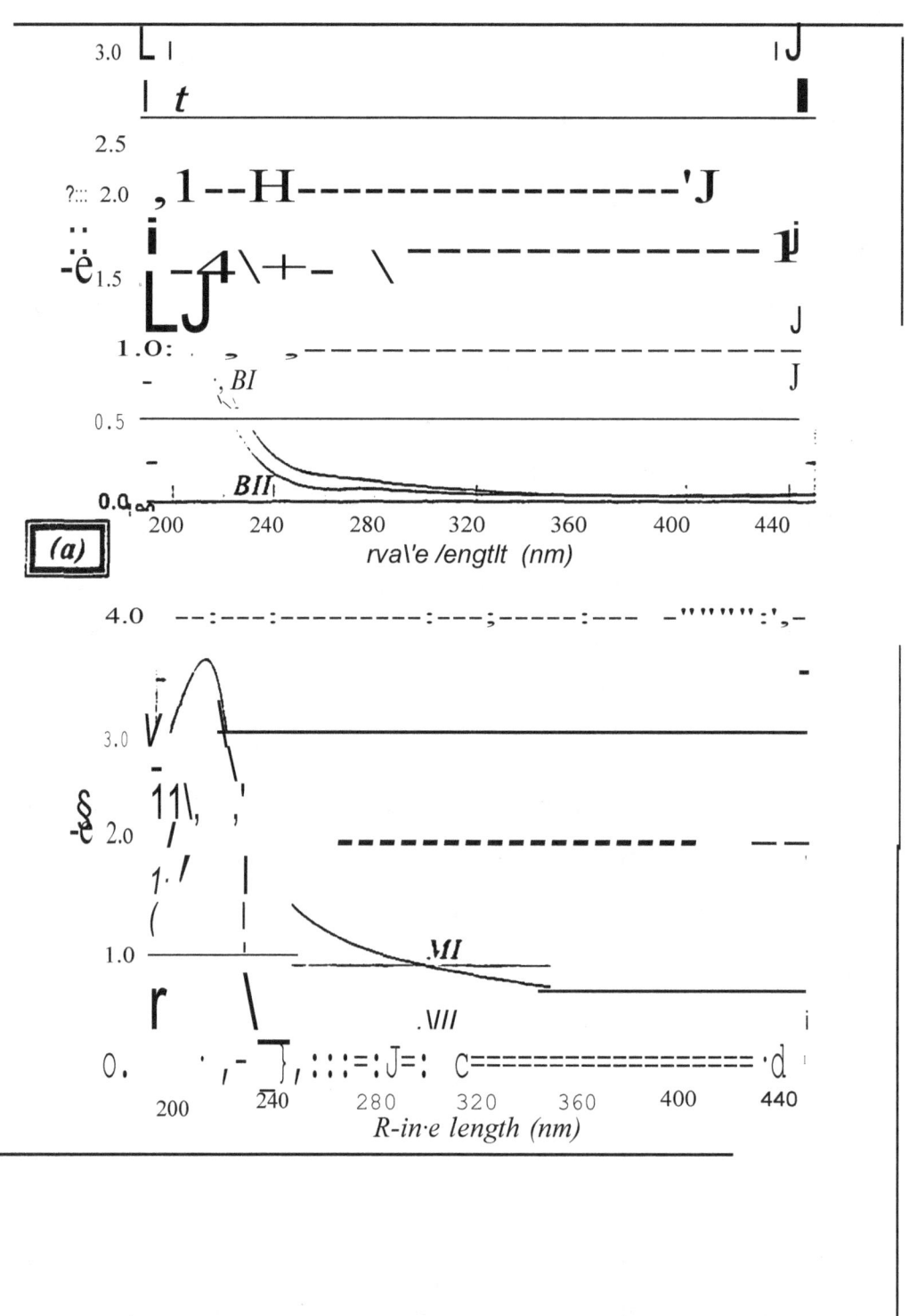

(a)

Figure (3.22) - The U.J- spectra of isolated estradiol receptors.
A.)Bl and B/1 receptors.
B):l-11 and .l·fll receptors.
All details are described in section (2.8.1).

3.7.2.1. pH effect

The pH of the solvent determines the ionization state of ionizable chromophores. Table (3.7) shows the *i-ma.x* values for estradiol receptors at different pH (2.7, 7.4 and 10.7). At an acidic pH 2.7, Bl-isolated receptor has two "-max values at 200.9 and 273.2run which were assigned to phenylalanine and tyrosine respectively. In BII-isolated receptor two Ama.x were obtained at 200.1 and 276nm. which were assigned to phenylalanine and tyrosine respectively. In fl-isolated receptor, one A-rna... was obtained at 201.1 nm, which was assigned to phenylalanine. In ·Mil-isolated receptor, two "-max wear obtained, the first one at :o2.3nm which was assigned to phenylalanine, the second at :276.:2nm was assigned to ryrosine.

At neutral pH 7.-+. BI and MI-isolated receptor, there were one Ama.x at :06.1 and :o9.6nm, respectiYdy. w·hich were assigned to phenylalanine. Ell-isolated receptor spectrum consists of two Ama.'. at :204.3 and 273..+ nm, \vhich were assigned to phenylalanine and tyrosine residues respectively. In l\I[II-isolated receptor three -max were optained at 202.8 and 260.3nm which were assigned to phenylalanine and at 223 .2nm which was assigned to tyrosine residues.

When the pH was increased from 7.4 to 10.7, there were no significant change in the Amax obtained for BI and :\-IT-isolated receptors. but has shown an increase in the Ama.x of tyrosine residues in BII and iVIII-isolated receptors. this result is due to the dissociation of the phenolic OH of tyrosine (pka = 10.07) giving an ionized form of this amino acid which absorps at higher wave length (red shift) 'b .

The spectral shifts of protein produced by pH cannot be simply attributed to the inductive effects of vicinal charges, such spectral changes must therefore be attributed mainly to rearrangements of secondary and tertiary structure,

---

although the possibility of field effects due to unusually close conjunction of charges to aromatic groups is not excluded [1169].

Table (3.7) - The effect of pH on the Amax of isolated estradiol receptors spectra. All details are described in section (2.8.2.1).

PH	BI-isolated receptor	BII-isolated receptor	MI-isolated receptor	MiI-isolated receptor
	"-max (nm)	Ama.T (nm)	AmtLT (nm)	Ama.T (nm)
2.7	200.9, 273.2	200.1, 276	201.1	202.3, 276.2
7.4	206.1	204.3, 273.4	209.6	202.8, 223.2, 260.3
10.7	:oi.2	205.8, 292.6	207.3	205.1, 278.2

3.7.2.2. Polarity effect on U.V. estradiol receptors spectra

The determination of whether an amino acid is internal or external by measuring the spectra of a protein in a polar and non-polar solvent is called the solvent perturbation method. In fact. proteins are rarely studied in completely non-polar solvents because must proteins are either insoluble or denatured in these solvents. However, significant solvent effects can be induced by use of a mixture of \Vater and a substance of reduced polarity such as ethylene glycoL ethanol, dimethyl sulfoxide and urea [11681] .

• *The effect of 20% ethylene glycol.*

Table (3.8) shows the effect of *20%* ethylene glycol at neutral pH. The Amax value of t:Tosine residues in BII and N'lii were shifted towards longer wavelengths tred shift) in *20%* ethylene glycol due to the hydrogen bonding of the OH groups oftyrosines with the solvent or with then-electron systemofthe benzene ring where tyrosine was functioned as a hydrogen donor, while the Amax value of phenylalanine in Bl. BII, rva and Mil was shifted towards shorter wavelengthst blue shift) in 20% ethylene glycol, this shift was attributed to n π· transitions.this blue shift in λma.x was accompanied with a decrease in the absorbancy of phenylalanine, this finding could be attributed to a change in the

protein stracrure, that phenylalanine residues were partly embedded in a hydrophobic region of the protein molecule.

- *The effect of 20% ethanol*

Table (3.8) shows the effect of 20% ethanol at nature pH. The i--max value of phenylalanine was shifted towards shorter wavelengths (blue shift) in 20o/o ethanol, this shift was attributed to π t rc· transitions. While the λmax value of t:Tosine was shifted towards longer wavelengths (red shift) in 20o/o ethanol due to the hydrogen bonding of the OH groups of Tosines with the solvent or with he ::-electron system of the benzene ring whc:-e r;·rosine \vas functioned as a hydrogen donort [168.,-C_J.]

These t\vo shifts m *l-m* ' were accompanied \Vith a decrease in the absorbancy ofphenylaianine and an increase in tr.e absorbancy oftyrosine. these findings could be .attributed to a chan e in the nrotein structure that brin the tyrosine residues to the surface of me protein \vhile phenylalanine residues \\·e:e pa.hly embedded in a hydrophobic region of the protein molecule.

The changes in the protein structure for BI-receptor may bring tyrosine residues to the molecule surface.

The change in 1-m x Yalue may indicate that the protein is sensitive to changes in the polarity of the soh·ent, \\·hich indicate that ccertain amino acid may be on the surface ofthe protein.

Table (3.8) - The effect of 20% ethanol, ethylene g(vcol and Di11SO on the 'AtrUlX of estradiol receptors spectra. All details are described in section (2.8.i.2).

Solvent	Bl- isolaJed receptor (nm)	BII- isolaJed receptor A-a (nm)	Ml- iso/aJed receptor Amax(nm)	MII- isolated receptor :i....-r(nm)
10°o et/zy/ene gll·col	205	103.4, 275.1	204.4	200.1. 226.4
20% ethanol	204.8. 224.1	203.4, 276.2	205.4, 223.8	200.2, 226
20%D. /SO	282.2	279.2	283.4	283.6
20% urea	219.6	216.4	217	113.1

- *The effect of 20% DMSO*

Table (3.8) shows the effect of DtvtSO on the estradiol receptors U.V. spectra at pH 7.4. Previous data in experiment (3.7.1) identified two peaks in BII and Mil-isolated receptor spectrum, these peaks were assigned to phenylalanine and tyrosine respectively, and one peak in BI and Ml-isolated receptor spectrum which was assigned to phenylalanine. In the presence of 20% DMSO pH 7.4, these two amino acids were buried inside the receptor molecules and a newer A_{max} was appeared for each isolated receptor, these are :282.:2, 279.2 283.4 and 2.83.6nm for Bl, BII. :\U. and Mil respectively which were assigned to tryptophan residues. The results indicate that DMSO effects the estradiol isolated receptors structurally. since many chromophores which were embedded in an interior region of the receptor molecule where they were inaccessible to the soh·ent came into contact \Vith it due to the unfolding of the molecule, and hence, different spectra \vere obtained [11701].

- *The effect of 20% urea*

Table (3.8) shows the effect of 20o/o urea at natural pH on the estradiol receptors spectra. In 20o/o urea. it was found that one $A_{ma.x}$ was obtained. For each receptor, at 219.6, :216.4.217 and 213.lnm for Bl, BII, Nll, and Mil respectively \vhich were assigned to tryptophan residues. The appearance of these new "-rna... Yalues indicates that the protein was defolded due to change in the secondary and tertiary structure of the protein that bring the tryptophan to expose to absorbance while phenylalanine and tyrosine residues were were buried inside the receptor molecules, also it was found that estradiol recepters are highly sensitive to change in the polarity of the solvent.

3.7.3. Spectropltototnetric pH titration of isolated estradiol receptors in benign and 1nalignant ovarian tu1nors

Spectrophotometric pH titration is the following of the change in absorbance of the chromophore with increasing pH (1 [68] >. Many studies of protein structure require the determination of pk values for proton dissociation from ionizable amino acid side chains, because these values give an indication of the location of the amino acid in the protein. This can often be done spectrophotometrically because dissociation often changes the spectrum of one of the chromophores, the observation of tyrosine dissociation was performed by measuring the absorption at 295nm 0-ma.x for the ionized form of tyrosine). and the obserYation of histidine dissociation was carried out by measunng the absorption at 211nm.

Figure (3.23 a and b) shows the pH titration curves of estradiol receptors for tyrosine and histidine respectiYely. (B) curves show that the pka values for tyrosine are 11, 11.5, 11.7 and 11 for BI, BIL 1v1I and !viTI-isolated receptors respectively while the pka values for histidine in (A) curves were equal to 6. 6.7, 6.5 and 7 for BI, BII, MI and Mil-isolated receptors respectively. From the same tigure, it was found that:

1) About 69A,63,59.6and 60.6% ofr;·rosine residues are located on the surface of the BI. BII, rvn and l\1111-isolated receptors molecule respectively.

2) About 30.6,37,40A and 39.4 of tyrosine residues are buried interior the folded snucture of the BI, BII. 11 and l\t1II-isolated receptors molecule respectiYely.

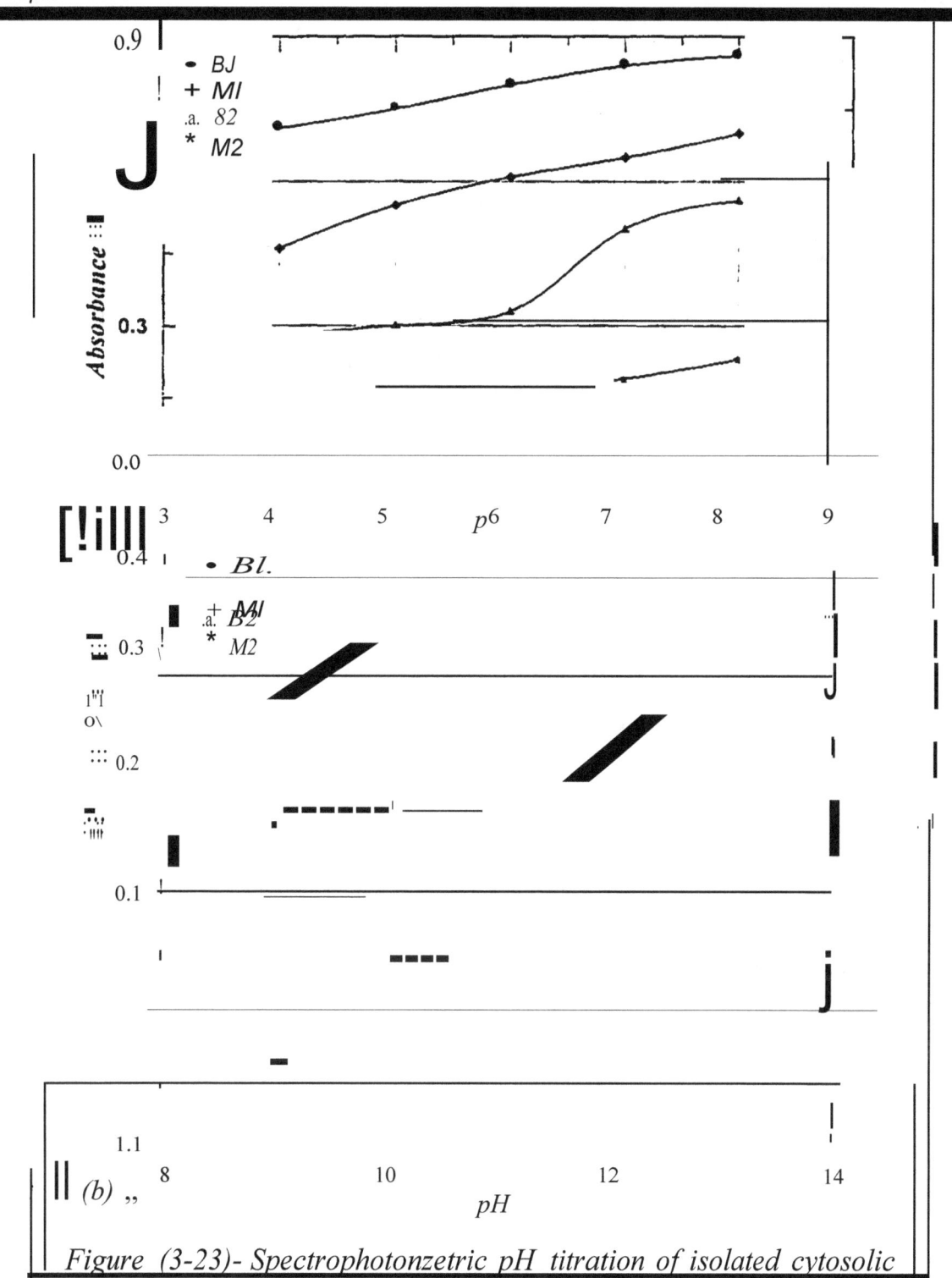

Figure (3-23)- Spectrophotonzetric pH titration of isolated cytosolic
·estradiol receptors for:
A) Histidine residues.
B) Tyrosine residues.

All details are described in section (2.8.3).

3) About 82.5,56,65. 7and 56.5 % of histidine residues are located on the surface of the BI, BII. l\t1I and l'vlii-isolated receptors molecule respectively.

4) About 17.5, 44, 34.3 and 43.5% of histidine residues are embedded in the LTlterior region of the BI, Bll, **rvn** and 1\t(.Il-isolated receptor; molecule respectively.

5) The internal tyrosine residues in BII, MI and :Mil-isolated receptors were in a strongly polar environment (e.g., atyrosine surrounded by carboxyl groups) while in BI-isolated receptor the surface of molecule and the internal tyrosine are in strongly nonpolar environment.

6) The histidine residues are largely present on the molecule surfa<::e ofBI and l\11-receptors and the internal residues are in a nonpolar environment whereas the internal histidine residue of BII and Mil-isolated receptors are likely to be in strongly polar environment.

7) The percent of external tyrosine residues in BI-isolated receptor was greater than that of:\111-isoiated receptor and the percent BII-receptor was greater than that of HI-receptor, on the other hand the percent of internal histidine in Nil-isolated receptor was greater than that ofBI-isolated receptor.

3.7.4. T/ze U.V. spectra of ^{125}/-estradiol and of the different ^{125}/-estradiol-receptors complexes

The binding of ligand (hormone) to the active site of a receptor frequently produces spectral changes in chromophores in or near the active site by affecting the polarity of the region or the accessibility to solvent this means that chromophores on the surface become inaccessible to the solvent by being buried in the region in which binding takes place or because a conformational change that buries or exposes a chromophore in another part of the molecule can accompany binding([1681].

Table (3.9) sho\vs the 1-ma.x values of ^{125}1-estradiol and its complexes with isolated receptors. The results of this study indicated that the binding of estradiol with its recc!ptors abolished the Ama.x values of free estradiol. and appearance of t)'Tosine residues on the molecule surface in all estradiol receptor types. The absorbance at 278.8 and 209.1nm obtained for estradiol may be attributed to the n 0'' 1t–)o ';'r: and n4 rc· electronic transitions respectively.

Table (3.9) - The i.max values of the C.V. spectra of ^{125}/-estradiol and its complexes with purified receptors. All details are described in section (2.8.4).

Isolated receptor	A-z(nm)
^{115}1-estradio/-R (BD	198
^{115}1-estradioi-R (BII)	197.3
^{15}∴ 1-estradiol-R (.14/)	199.2
∴ /-estradioi-R (.l-f//) ^{15}	198.4
^{115}1-t!Stradiol	278.8, 209.1

Conclusions

1) The modification protocol for the assay of estradiol receptors is capable of analyzing these receptors and the procedure is suitable for the assessment of estradiol receptors in benign and malignant ovarian tumors.

2) _-\ higher incidence of estradiol receptors was obtained in malignant than in benign o'\·arian tumors: therefore the malignant rumors were more estradiol dependent than those of benign rumors.

3) The kinetic studies of the $^{1\,5}$1-estradiol binding to its. receptors showed that :he binding reaction is a temperature and time depe:1ded process.

4) The association of estndiol with its crude benign and malignant ovarian n;mor rece::-tors were spontaneously occur (Gc < 01 and the binding reactions were entropically dri\·en C: S:: > 0).

5) The spectroscopic studies on isolated estradiol receptor re\·ealed a characteristic spectrum for each receptor.

Future work

1) Evaluation of the biological activity of estradiol receptors in benign and malignant ovarian tumors.

2) Evaluation of the estradiol level m unnary and salivary in patients with ovarian tumors.

3) Purification of ovarian estradiol receptors in different subgroups of tumors and using different technique in purification.

4) Molecular characterization of those receptors of ovarian tumor.

5) Spectroscopical characterization of purified estradiol receptors by I.R.. N.M.R.. fluorescence. x-ray diffraction analysis.

References

1. Kaplan L.A. and Pesce A. J., *"Clinical Cllemistry, Theory, Analysis and Correlation··*, 2nd ed., The C.V. Mosby Company; 1989, pp. 611,612,651-657.

2. Davidson V. L., and Sittman D. B., *"Biochemistry"*, 4th.ed, Lippincott William and Wilkins, 1999, pp. 112, 113.

3. Ryan K. J., Berkowitz R. and Barbieri R. L., ••*Kistllers Gyllecology, Principles and Practice"*, Sth.ed, Year Book Medical publishers, 1990, pp. 23, 24, 264-268.

4. Caret R. L.. Denniston K. J.. and Topping J. J., *"Principles and Applications of Inorganic, Organic, and Biological C/temistry"*, 2nd.ed, Times Mirror Higher Education Group, Inc.. 1997.p.391.

5. Yen S. C., and JatTe R. B.. ··*Reproductive ndocrinology"*, 2nd.ed, W.B. Saunders Company, 1986. pp.276, 280-285.

6. \tlcnatty K. P.. lakris A.. Degrazia .C. Osathanondh .R.. and ryan K. J., *"J.Ciin. Elldocrillol. 1\tfetab."*. 1979;49(5): 687.

7. King R. J.. and Mainwaring W. 1., *"Steroid-Ce/1/nteractiolls"*. Buttenvorth and Co. Ltd., 1974, p.2.

8. Porth C. M., *"Patlzoplzysiology"*, 4th.ed.• J. B. Lippincott Company. 1994, pp. 750,751.

9. Green B. and Leak R. E.. ··*Steroid Hormones, A Practical Approach"*. IRL Press Limited; 1987. p. 2.

10. Telleria C. L Zhong L.. DebS., Srivastava R K.. Park K. S.. Sugino N.. Park 0. K.. and Gibori G., ••*E,docrillology"*. 1998:139(5):2432.

11. Hubl W., ··*Hormone Diagnosis for Fertility Disorders"*, Dresden-Friedrichstadt Hospital, 1992. p.2.

12. Guyton A. C.. and Hall J.E.. •·*Text Book of Medical Physiology"*, 9th.ed., W. B. Saunders Company, 1996, pp.1022.1023.

13. Jubiz w., *'"Endocrinology"*, 2nd.ed, The McGraw-Hill Book Company, 1987,p.439.

14. Burtis C. A.. and Ashwood E. R., *"Tietz Text Book ofClillica/ Cltemistery"*, 3rd.ed, W. B. Saunders Company, 1999,pp. 1608-1610.

15. Pardridge W. M., *"C/in. Endo..Wetab."*. 1986: 15:259.

16. Lehninger :\. L.. *··Biocltemistry"*. Wonh P:1blisher C-.:C. ICS.-\). 1982.pp.74. 45.

1*i*. Greenspan F. S.. and GardnerD. G...*Basic and Clinical Endocrinology"*. \fcGrav.- Hiil Company. 2001, pp. 66.

18. Hay \\·. \V., HaY\\"Ord A. R... Levin :YL J.. and Sondheimer J. M.• *"Current Pediatric Diagnosis and Treatmellt"*, 1-+th.ed. Appleton and lange. 1999. pp. 60 l. 607. 608.

19. :\tlurray R. K.• Granner D. K.. y[ayes P. A. and Rodwell V. W.• *"'Harper's Biochemistry"*; Twenty Seconded.: A Lange yfedical Book: 1993; ppA67. -1-68.

20. Adashi E. Y., *"Reproductive Endocrinology,..* \\·. B. Saunders Co. 1992. pp.l81.

21. Carr B. R.. *"'Williams Textbook of Endocrinology"*. 8th.ed.. \V.B. Saunders Co. 1992. pp.733.

22. Pennington G. \V.. and aik S., *"Hormone Ana(t:sis .Uetlrodology and Clinical Interpretation"*, Volumeii. CRC Press.lnc.. 1981.pp. *33.35*.

·ᵁ·ᴸ Lloyd C. \V.. Lobotsky L Baird D. T.. :\1cCracken J. - --Weiss J.. Pupkin :\L. Zanarru J..and P...rgaJ., *"J.C/in.Endocrinot:·*. :\.let..1971:32:155.

2 . Baird D. T.. Burger P. E.. HeaYon-Jones C. D. and Searamuzzi R. T.. *··J. Endocrino/"*. 1974; 63: .:!01.

...::;. 8ecᵏmann R. B., an&ⁱ Lᴵₘg f. \⎯.. *"'Obstetrɾ'cs anᵈ G) •neco/ogy•".")*nᵈ.-"ᵈ..1...0 1ᵗ·⁻ H.ᵞp. 461.

26. Erickson G. f., *"Clin. Obstet. Gynaeco/."*. :978:.:!1(1):31

2-. lalle:- B.. and Stron C. .-\... ..*Reproductive EJrdocrinolog)·'·*. \\.*i*. B. Saunders Co. 1991,pp.156.

28. Carr B. R., and Blakwell R. E.. *"Te:abook of Reproductive 1Uedicine...* 2nd.ed.. Appleton and Lange. 1998. p.223.

29. Johnson L. R.. *"Essemia/.llt!dical Plrysiology"*. :!nd.<!d.. Lippincon-Ra\·l!n Publishers. 1998. P.650

30. Heinonen P. K.• \.lorsky P...-\ine R.. Koi\ula T.. and Pystynen P., ..*llaturitas"*. 1988;9:.325

31. James\". H.. Reed \.L J.. and Folkerd E. J.. ••*J. Steroid Biocltem."*. 1981:15:235.

32. Reed M. J.. Beranek P. A.• Ghilchik M. W., and James V. H.• ""*Obstet. Gynecol*", 1985; 66:361.

33. Welshans W. V., Lieberman M. E.. and Gorski J., "*Nature*", 1984; 307(23): 747.

3-J. Jensen E.V .• and DesombreE.R., "*Science*", 1973;182:126.

35. Voet D., and Voet J.G., "*Biocllemistery*", John Wiley and Sons, 1990, p.1026.

36. Gold J. J., and Josimovich J. B., "'*Gynecologic Endocrinology*", 3rd.ed. Harper and row Publishers INC., 1980. pp. 97-99.

37. Joab L Radanyi C.. Renoir M.. Buchou T., Catelli M.• Binart N., MesterJ., and Baulieu E.. ""*Nature*". 1984:308(26):850.

38. King W. Land Greene G. L., "*Nature*", 1984: 307 (23): 745.

39. Smith E. L., Hill R. L.. Lehman I. R.. Lefkowitz R. J.. Handler P.. and White A.. "*Principles of Biochemistry: Jfamma/ian Biochemistry*". 7th.ed.. McGraw-Hill Book Company. 1985, P.390

40. Berne R. M., and L vy M. N., "*Physiology*", 3rd.ed. Mosoby Year Book. 1993. pp. 1004. 1012.

.Jl. Hart D. M., and t\orman J.. "*Gynecology Illustrated*". Sth.ed. Harcoun Publishers Limited. 2000, pp.54. 254.

42. Erickson G. F., "*C/in. Obstet. Gynecol*.... 1978;21 (1):31.

43. Seeley R. R., Stephens T. D.. and Tate P...*A11atomy and Physiology·*•. 2nd. ed.. mosby-year Book. Inc., 1992. P. 932

4-t. Brothenon J., ··*se...: Hormo11e Pharmacology*", Academic Press. Inc. 1976. p. 139.

45. Murray R. K., Granner D. K.. Mayes P. A.. and Rodwell V. W.. "*Harpers Biochemistry*", 25th.ed.. Appleton and Lange, 2000, pp. 603,604.

46. Kannan C. R., "*Essentil Endocri11ology*", Plenum Medical Book Company, 1986. p.328.

.J7. Bychkov V., and Isaacs J. H.. "'*Pathology in tlte Practice ofGJmecology*". Mosby-Year Book, INC., 1995, pp. 272-275.

-'8. Schneider A. S, and Szanto P. A., "'*Pathology*", Mass Publishing Co., 1997, pp. 287-

49. Cotran R. C.. Kumor V., and Collins T.• '"*Robbins Patltologic Basis of Disease*". 6th.ed. W. B. Saunders Company, 1999,pp. 1068-1070.

50. Daniel L., Dawood M. Y.. and Hammond C. B., "*Greens Gynecology: Essentials of Clinical Practice*", 4th.ed. Linle Br0\\-11 and Company, 1990. pp. 531-540.

51. Jeffcoate W.. "*Lecture Notes on Endocrinology*". Sth.ed. Blackwell Scientific Publications. 1993. p. 209.

52. Benson R. C.. and Pemoll t. L., "*Hand Book of Obstetrics and Gynecology*". 9th.ed. 1994, pp. 562-572.

53. Nnlambron N. C.. Mprse .-\. N.. and Wallacl E. E....*Tite Jolrns Hopkins ,\.1anual of Gynecology and Obstetrics*". Lippincn Wilhiams and \\"ilkins, 1999. p.370.

54. Iscovich L Shushan A., Scheuker J. G.. and Paltiel 0.. ··*cancer*", 1998:82(1): 147.

55. Gotlieb W. H., Flikker S.. Devidson 8.. Korach Y.. Kopolovic J.. and Baruch G.. ••*cancer*", 1998; 82(1): 141.

56. Whittield C. R.. •·*Dewlzursts Textbook of Obstetrics am/ Gy1raeco/ogy*". 5th.ed.. Blackwell Science Ltd., 1995.p.759.

57. Pollock R. E.. -*Jl1anual ofCiinical Oncology*", 7th.ed.\Viley-liss.INC.. 1999. pp. 543-547.

58. Zanena G..Bonazzi C., Cantu M.G.• Bini S.. Locatelli.-\.. Bratina G..and Mangioni C.. "*J. Clilr. Oncol.*", .:!00 1;19(4): 1015.

59. Bland K. I.. Daly J. M., and Karakousis C. P., ··*surgical 011cology: Contemprary Principles and Practice*". The :YlcGraw-Hill Company. :::001. p.991.

60. Gunderson L. L.. and Tepper J. E.. ..*Clinical Radiation Oncology*". Churchill Livingstone, :WOO. pp. 940-943. 949.

61. Nagasawa H.. ""*Hormone Related Tumors*", Japan Scientific Societies Press. 1981.p..:!43.

62. Slonnan B. J..and Rao B. R....*Cancer Letters*". 1989:45:213

63. Menzin A. \V.. Loret J. R.. Bilker W. B.. wheeler J. E.. Rubin S.C.. and Feinberg R. F.. '"*Cancer*", 1998; 82(1): 152.

64. Neijt J. R., "*N.Engl. J. Med.* ". 1996; 334(1): 50.

65. Heinonen P. K.• """.Waturitas". 1991: 13: 117.

66. Walker G. R., Schlesselman J. J.. and ess R. B..-Am. J. Obstet. GynecoL ". 1001: 186(11: 8.

67. Nieto J. J., Rolfe K. L Maclean A. B.. and Hardiman P.. ""T/te Lancet", 1999; 354: 649.

68. Lumbiganon P., ;"Contraception" 1994:49: 103.

69. Westhoff C., Britton J. A.. Ganunon :\1. D.. \Vright T.. and Kelsy J. L.. ··Am. J. Epidemiol", 2000: 152(3): 141.

70. Joly D.- Lilienfeld A. M.. Diamond E. L.. and Bross I. D.. ··Am.J. Epidemiol.-. 1974: 99:190.

71. Ness R. B.. Grisso J. A.• and Klapper J•• "Am.J. Epidemiol. ". :!000: 152(3): 233

72. WeS""Jloif C.• Britton J. A.. Gammon :\-1. D.. \Vright T.. and Kelsey J. L.. ···.-tm. J. EpidJmi.ol", 2000: 152(3):242.

73. Buns C.. and Freedman R. S.. ··Cancer Control", 1999: 6c4): 335.

74. Scon J. R.. Disaia P. J.. Hammond C. B.. and Spellacy \V. -. "Danforths Obstetrics a11d Gynecology". 8th.ed., Lippincott Williams and Wilkins.1999.p.678.

'75. Edmonds K., "Dewlmrsts Text Book of Obstetrics a11d Gy11aecology for Postgraduates", 6th.ed. Black \Veil Science LTD. 1999. pp. 590-594. 599. 600.

76. Schild.kraut J. M.. Collins X. K.. Dent G. A.. Tucked. A..• Barren J. C.. Berchuch A.. and &yd J.. '"Am.J. Obstec. GynecoL ". 1995: 127(3): 908.

77. Frank T. S., '"Cancer Control". 1999. 6 4): 327.

78. Tiernyy L. M.. lcphee S. L and Papadakis :\-L A.. "Currellti'Jtledical Diagnosis and Trea nt", 33th.ed. Appleton and Lange. 1994. p. 601.

79. Markman M., "J. Cancer Res. Clin. Oncor", 1994: 120:257.

80. Newcomb E. \V.. Sosnow M.. Demopoulos R. 1.. Jacquone A. Z.. Sorich J.. and Speyer J. L. -.-tm. J. Pathof. ", 1999:154t1): 119.

81. Marks J. R., Davidoff A. L Kl!ms B. J.• Humphrey P. A.. Pencej. C.. DodgeR. K.. Clarke-Pearson D. L.. lgiehan J. D.. Bast R. C.. and .Berchuck A.. ··Cancer Res.". 1991: 5\:2979.

82. \\"hittemore A. S.• and Paffenbarger R. S., '"Am.J. Epidemiol ... 1988: 128:1228.

83. Cramer D. W.. Welch W. R.. Sclly R. E.• and Wojciechowski C. A.• ··*cancer*", 1982: 50:372.

84. Roa F. J., *"Lancet"*, 1978; 2:744.

85. Cavalli F., Hansen H. H.• and KayeS. B., ·*'Text Book ofJJedical Oncology"*, Martin Dunitz Ltd, 1997, pp. 88, 92.

86. Richardson G. S., Scully R. E. 1\ikrui N., and Nelson J. J., *"N.Engl. Jllled."*, 1985; 321:415.

87. Casciato D. A., and Lowitz B. B., •·*,J.fanual of Clinical Oncology"*, Lippincott Williams and Wilkins. 2000. pp. 257. 258.

88. KentS. W., and Me Kay D. G.. *'"Am. J. Obstet. Gynecol"*, 1960; 80:430.

89. Piver M.S.. Lele S.. and Barlow J. J.. ··*Obstet Gynecol"*. 1976:48:312.

90. Nnktnnedy C. R., and Gordon H.• ••*Br. J. Obstet. Gynecot:*·. 1981; 88:1186.

91. KentS. W.. and Me Kay D. G., *"Am.J. Obstet. Gyllecol"*. 1960; 80:430.

92. Meire H. 8.. Farrant P. and Guha T., *"Br. J. Obstet. Gynacol."*, 1978:85:893.

93. Eichner E., and BoveE. R.. *"Obstet. Gynecol."*. 1954; 3:287.

94. Parker B. R.. Castellino R. A., Fuks Z. Y.• and Bagshaw M. A., *"Cancer"*. 197-k 34:100.

95. Chang T. C.. Jain S.. Hsueh S.. Tsai C. S.. Chen H. L.. and Chang C. N.. ··*J. Reprod11ctive .\4ed."*, 2001: 6(3): 267.

96. Larsen J. F.. Pedersen 0. D.. and Gregersen E.. *"Acta Ohstet. Gynecol. Scand."*. 1986;65: 539.

·97. Ehabbakh G. H., and Kaiser J. R., ..*J. Reproductive ,\-fed.*·. 2000: 45(3): 231.

98. Tierney L. M., Mcphee S. *1.,* and Papadakis M. S., *"Current l'vledical Diagnosis and Treatment"*. 40th ed, The McGraw-Hill Companies, Inc.. 2001, pp. 746,

99. Nagele F.. and Magos A. L. *"Am.J. Obstet.Gyneco/."*, 1996; 175:1377.

100. Studd J., ·*'Progress in Obstetrics and Gynecology"*, Vol.9, Longman Group U .K Limited, 1991, pp.358-360.

101. Emoto .\IL Lwasaki H.. Mimura K.. Kawarabayashi T.. and Kikuchi M., *'"Cancer",* 1997; 80(5) :899

102. Pandha H. S., Waxman J.• and Sikorak, *'"Br. J. Hospital Wedicine",* 1994; 51(6) :297

103. Burtis C.A., and Ahwood E.R.• *"Tietz Text Book of Clinical Cllemistry",* 3rd.ed., W. B. Saunders Company, 1999, p. 723.

104. Van agell J. R.• Donaldosn E. S.. Hanson M. B.. Gay E. C., and Pavlik E. J.. *"Cancer".* 1981. 48: 495.

105. Nouwen E. J., Pollet D. E., Schelstraete J. B., Eerdekens M.\V., Hansch C., Van de
\
Voorde A.. and De Brae M.E.. *"Cancer Res.",* 1985; 45: 892.

106. Verdon B.. Berger E. G.. Salchli S.. Goldhirsch A.. and Gerber A.. *""Clill. Chem.",* 1983;29:1928. •

107. Kuwashima Y., Uehara T., Kishi K.. Shiromizu K., Matsuzawa M., and Takayama S., *"J. Cancer Res. Clin. 011co/.",* 1994: 120 :672.

108. Kobawat S. E.. Bast R. C., Welch \V.R, Knapp R. C.. and Colen R. B., *"Am.J. Clin. Pati10L ".* 1983;79:98.

109. Tobias J. S.. and Griffiths C. T., *"N. Eng/J. iUed.",* 1975: 298:818.

110. Van 'Sagell J. R.. Donaldson E. S.. Wood E. G.. and Goldenberg D. M., *··ca1 1cer",* 1978: -1-:!:1527.

111. Mioni S.. Agnannos S.. Canevari S.. Diotti A.. Orlandi R.. S01mino S.. and Colnaghi :Yl. I., *"Cancer Res.",* 1985: 45: 826.

112. Tagliabue E.. Menard S.. TotTe G. D.. Barbanti P.. Mariani R.. Porro G., Colanaghi :YL I., *"Cancer Res.",* 1985: 4-5: 379.

113. Kuoppala T.• Heinola M., Aine R.. Isola J.. and Heinonen P. K., *"Il1tJ. Gy1 1ecol Cancer",* 1996; 6: 302.

114. Einhorn N., Bast R. C.. and Knapp R. C., *"Obstet. Gy1 1ecol. ",* 1986: 67:414.

115. Bast R. C.• Kulg T. L.. Jolm E.• Jenison E.. and NilotT J. M.. *••N. Ellgl. J. 1Wed. ",* 1983: 309: 883.

116. Geisinger K. R., Kute T. E.. Pettenat M. J., \Velander C. E.. Dennard Y.• Collins L.A.. and Berens M. E., *••ca1 1cer",* 1989: 63(2): 280.

117. Rustin J. S.. Nelstrop A. E.. Bentzen S. M.. Bond S. J.• and Me Clean P., *"J. C/in. Oncol. "*, 2000: 18(8): 1733.

11 . Van Nagell J. R., Donaldson E. S., Hanson M. B., Gay E. C., and Pavlik E. L *··Cancer"*, 1981; 48:495.

119. Di Saia P. J., Marrow C. P., Haverback H. Land Dyce B. J., *"Ca1 1cer"*, 1977:39:2365.

120. Vassilomanolakis M.. Koumakis G., Barbounis V.. Hajichristou H.. Tsousis S.. and Efremipis A.. *'"Oncology"*. 1997: 45: 199.

121. Geisinger K. R., Berens M. E.• Ducken Y., Morgan T. M, Kute T. E.. and Welander C. E.. *·"Cancer"*. 1999; 65(5): 1055.

122. S1otman B. J., Baak P. A.. and Rao B. R.. *"Eur. J. Obstet. Gynecol. "*. 1990: 38:221.

123. Cannina E.. Ditkoff E. C.. Malizia G.. Vijod A. G.. Janni A.. and LaboR. A., *"Am. J. Obstet. Gynecol. "*, 1992; 167(6): 1819.

124. Greenblatt R. B., Colle M. L., and :Vfahesh V. B., *"Obstet. Gynecol. "*. 1976: 47:383.

125. k.-uhnel R.. De Graaff. J., Rao B. R.. and Stolk J. G.. *"·J. Steroid Bioclrem. "*, 1987;26:393.

126. Slotman B. J.. and Rao B. R.. *'"The Cancer Journal"*. 1989: 2: 373.

127. Frod L. C.. Berek J. S.. and Iagasse L. D.. *""Gynecol. Oncol."*. 1983:15:299.

128. Kauppila A.• Vierikko P.. Kivimem S.. Stemback F.. and Vihko R.. *--obstet Gy11ecol. "*. 1983:61:320.

129. Lanne M.. *"Acta Obstet Gynecol Scand ...* 1984: 63: -+97.

130. Bergqvist A., Kullander S.. and Thorell J.. *"Acta Obstet Gynecol Sca1 1d'·*. 1981: 101: *75*.

131. Abrahamsson. G., Janson P.O.. and Kullander S., *'·Acta Obstet Gynecol Scand"*, 1990: 69: 527.

132. Heinonen P. K., Tuimala R.. Pyykko K.. and Pystynen P.. *"Br. J. Obstet Gy11ecor·*. 1982: 89: 84.

133. Heinonen P. K., Koivu1a T.. Rajaniemi H.. and Pystynen P., *"Gy11ecol Oncol"*, 1986; 25: 1.

134. Aune R. M.. and Noel W.. *"La1 1cet"*. 2001: 358: 438.

1 3 Deutscher :\-1. P.. *"i'Wetlrods in Enzymology"*, Volume 182: Academic Press Inc.: 1990: pp.197

136. Lowry 0. H.. Rosebrough N.J.. FarrA. L. and Randall R. J.. *"J.Bioi. Clzem:·*, 1951; 193:265.

137. Clinical Assay. *·'Gamma Coat Estradiol-/ RIA Kit"*, Immunotech International (FRANCE).

138. lorris B.J.. *··Clinical Chemica A eta"*, 1976. 73: 213.

139. Scatchard G.. *""Ann. N.Y. Acad. Sci."*, 1949.51:660.

140. Segall. H.. *"Biochemical Calculations"*, 2nd ed..John Wiely and Sons. 1976. p.311

141. Scopes R.K.. *··Protein Purification"*, Principles and Practice: 2nd ed.: Springer-verlag: 1987; pp. 196-198.

142. ThompsonS. A.. Johnson !v1. P.. and BrookS. *C.:, '"Tire Prostate"*. 1982; 3:45.

143. Heinonen P. K., Koh.rula T.. Rajaniemi H.. and Pystynen P.. *"Gynecol. Oncol."*. 1986;25: I.

144. Habib F. K.. Lee I. R.. Stitch S. R. and Smith P. H. *"J.Endocri"*. 1976: 71: 99.

145. Rao B. R.. Slotman B. J..Geldof A. A.. and Dinjens *N.* M., *"lilt. J.Gynecol. Pat/to/."*. 1999;9(1):47.

146. Lineweaver H.. and Burk D., *"J.Ann. Chem. Soc."*.-1934:56:658.

147. Shiu P.R. C. and Friesen H. G.. *··Bioclzem. J."*. 1974. 140: 310.

1-'8. Horo L.S. and Talaments F.G.. *··.Hoi. Cell Endocrilzol."*. 1985. 43: 199.

149. Al-Samraee I. H., *".Ho/ecular Cltaracteri:ation of Estradiol Receptors in Uterine Elldometrium Effected by Benign and JJa/ignmzt Tumors"*. 1997,M.Sc.. thesis. supervised by Al-Mudhaffar S.A.. College of Science. Baghdad University.

150. Daxembichler G.. Grill H. J.. Wiesinger H.. Winliff J. L.. and Dapunt 0., *"J,fultiple .lfolecular Forms of Steroid Hormone Receptors"*. A garwal M.K. editor. Elsevier. North Holland Biomedical Press. 1977. p. 163.

151. Melander W. and Horvath C.. *"Arch. BiochenL Biophysi."*, 1977; 183: 200.

152. \Valsh M.P.. Vallet B.. and Autric F., *"J.Biol.Cizem.""*. 1979, 244: 12136.

153. Liv.;ack G., ""*Biochemical Actions of Hormones*": Vol.IV. Academis press: 1977, p. 371.

154. Leake A.• Chrisholm G. D.. Busuttil A. and Habib F. K.. ·"*Acta Endocr.* ", 1984; 105(2): 281.

155. Joan Reed L and Stitch S. R.. ·•*J. Endocri*", 1973: 58: 405.

156. Kadhum M, "*Evaluation of Some Biochemical Constituents (Enzymes and Trace Elements) in Breast Tumor Patients*". 1996, Ph.D. thesis, superYised by Al-:\1udhaffar S.A.... College of Science. Baghdad University.

157. Cooper J. R.... Bloon F. E.. and Roth R. H. "*Biochemical Basis of Neuropharmacology*", Oxford Cniversity Press Inc.. 1978. p. 60.

158. Liao S., Liang T... Fang S.. Castaneda E. and Shao T.C.. .*J. Bioi. Clzem.*". 1973: 248i17): 6154.

159. Sloanan B. J.. auta J.P., and Rao B. R.. -*cancer*". 1990: 66{4): 740.

160. Slonnan B. *1*.. Kultnel R.. Rao B. R.. Dukhuizen G. H.. Graaff J.D.. and Stolk J. G.. •·*Gynecol. OncoL* ., 1989;33:76.

161. \\'eiland G.˗\. and :Vlolinoff P. B., "*Life Science*". 1981: 29: 31 .

162. l'emethy G.. and Scheraga H..-\., "*J. Phys. Clzem.*", 1962. 66: 1773.

163. \\-alebroeck L Van Obberg.hen E.. and De le:1s P.. "*J. Bioi. Clre11L*"- 1979, 25-k *7736*.

164. Kauzmann \\-.. -*Adv. Prot. Chem.*", 1959: 14: 1.

165. Ross P. D.. and Subramanian S.. "*Biochem:·*. 1981: 20: .3096.

166. Blumenthal D. K.. and Stull J. T.. "*Biochem:·*. 198 : 21:2386.

167. Lapone D. C.. "\\"iennan B. Land Storm D. R.. ""*Bioc/zern.*", 1980: 19: 3814.

168. Freifelder D.. .*Pizysica/ Biochemistry'*", 2nd ed..: 1982: pp. 500-503. 511.

169. Yanari S., and Bovey F. A.. --*J. Bioi. Chem.*", 1960: 135(10): 1818.

170. Lc!'ach S. J., and Scheraga H. - -- "*J. BioL Citem.*", 1960: 235(10): 2827.

www.ingramcontent.com/pod-product-compliance
Lightning Source LLC
Chambersburg PA
CBHW080812180526
45168CB00006B/2421